JIT/Lean Methods and Japanese Management

JIT/Lean Methods and Japanese Management

An integrated approach to the study of the Japanese manufacturing company

Lumbidi KUPANHY

iUniverse, Inc.
New York Lincoln Shanghai

JIT/Lean Methods and Japanese Management
An integrated approach to the study of the Japanese manufacturing company

iUniverse books may be ordered through booksellers or by contacting:

iUniverse
2021 Pine Lake Road, Suite 100
Lincoln, NE 68512
www.iuniverse.com
1-800-Authors (1-800-288-4677)

Because of the dynamic nature of the Internet, any Web addresses or links contained in this book may have changed since publication and may no longer be valid.

The views expressed in this work are solely those of the author and do not necessarily reflect the views of the publisher, and the publisher hereby disclaims any responsibility for them.

ISBN: 978-0-595-45438-9 (pbk)
ISBN: 978-0-595-89750-6 (ebk)

Printed in the United States of America

This book is dedicated

To
My family

To
The memory of
My beloved parents
Whose precious blood
Like an eternal stream
flows Inside me;
Whose blood like a hot spring
keeps me warm.

Bigraphical notice

Professor Lumbidi KUPANHY teaches Operations Management, Logistics and Supply Chain Management at Euromed Marseille Ecole de Management, France. After getting his Ph.D. from Osaka City University, he worked for a private company before becoming a faculty member of Wakayama University, Japan. He left Wakayama University to join Euromed Marseille in 2003.

His research areas include Japanese production system, lean supply chains and Japanese management style. He has extensively published in these areas. His teaching experience includes visiting professorship at INSEAD, Leipzig School of Management (HHL, Germany), Jia Tong University Antai Business School (Shanghai, China), Faculté des Sciences Economiques, Marseille Colbert (Université d' Aix-Marseille).

He is the scientific and academic coordinator of two Master Programs: International Master in Management (Operations & Supply Management Cluster) and Maritime Management & International Logistics.

Contents

List of Illustrations

List of Tables

Acknowledgements

This book would not have been completed without the help, support, generosity and contribution of many people and institutions.

I feel indebted to the Chamber of Commerce and Industry of Osaka, Japan, for having provided me with ready-to-mail addresses of its company members belonging to the sector of manufacturing; and to the companies that kindly returned to me a properly filled-in questionnaire. I can not forget my Japanese students at Wakayama University who, ten years later, collected data in Wakayama Prefecture using exactly the same questionnaire.

I would like to thank the management of the following companies/plants for having given me a chance to visit their respective production sites: Daihatsu (Shiga), Daikin (Sakai), Hirotek (Hiroshima), Honda (Suzuka, Mie), Kao (Wakayama), Kawasaki Heavy Industry (Kobe), Kubota (Sakai), Mazda (Hiroshima), Mitsubishi Cool (Wakayama), Mitsubishi Denki (Kobe), Mitsubishi Motor Corporation (Mizushima, Okayama), Mori Seiki (Mie), NKK (Fukuyama Works), Nissan (Tochigi), Noritsu Koki (Wakayama), Shima Seiki (Wakayama), Tekuno Plaza (Hiroshima), The Japanese Mint Bureau (Osaka), Toyota (Toyota Shi, Aichi), Tsuneishi Ship Building Co. (Hiroshima)

Besides, I would like to thank Takasho Co., Ltd. (Kainan) and Wakayama University who gave me the opportunity to experience first hand the Japanese management style at a private, for-profit company (Takasho); and at an academic institution (Wakayama University). President N. Takaoka, Managing Director Hiramatsu, Professor T. Kojima, Professor A. Oda, Professor A. Takeuchi deserve all a special and particular mention of gratitude.

I owe so much to the Graduate School of Business of Osaka City University who provided me the funds I needed in order to continue my field research and increase the sample size. I am very indebted to the late professor S. Uemura, and Professor H. Kameda for all they did for me during my research at Osaka City University.

I have to thank also my family for its kind support without which it would have been difficult to complete this book.

Introduction

0.1. Object and scope of the study

The object of the present study is to try to understand Japanese manufacturing methods and their immediate surrounding environment. The focus is therefore on the production techniques as well as the manufacturing and management strategies of the Japanese company.

As everyone knows, external forces more than internal ones often constrain companies to review their strategies and/or organization[1]. Concerning Japanese corporations, they have been shifting, since the end of the Second World War, their management strategies, production techniques and their organization structures in response to successive and various environment changes.[2]

The study is going to focus on some features of the Japanese corporation that can be qualified as constants of the Japanese management. They are common to Japanese companies in general and most of them are very closely related to the human resource management. The production method that is said to lie at the base of the competitive edge or advantage of the Japanese manufacturing company is called just-in-time (JIT). The JIT was initiated and developed at and by Toyota. In 1990, the JIT production was recognized as a lean production system[3]. Since then, lean and JIT systems have been considered as conveying the same meaning. They are synonymous and interchangeable. In other words, JIT is now known and referred to as the lean production. One should not lose sight of the fact that, besides being a set of production methods, JIT can be viewed as a production strategy.

---

1   Although external forces such as environment constraints press for changes in organization and strategies of business enterprises, internal forces sometimes, even most often, tend to resist such critical changes due to their inertia.

2   Sh. Uemura (1993, p. 92) summarizes in a table describing the evolution of different economical environment changes (1945 to 1990) of Japan, the corresponding management strategies and the implied changes in organization structure.

3   J. P. Womack, T. Daniel and D. Roos, *The machine that changed the world*. The Story of lean production, 1990, Rawson Associate, 1992

Figure 0–1 Company representation in terms of hardware and software

Hardware	• Automation • Plant and Equipment • Computerization • Energy Utilization
Software	• Employment Practices • Production Methods • Decision Making • Information Diffusion • Social Innovation

Source: Adapted from Ch. J: McMillan, 1985, p. 27

An attempt of a bipolar representation of the Japanese company adapted from McMillan[4] in terms of hardware and software is shown in Figure 0–1. The hardware is made up of features resulting from successive strategic changes whereas the software can be viewed as especially consisting of common and constant characteristics of Japanese companies. The software consists of employment practices such as decision-making process, information diffusion, production methods, and social innovation[5]. Social innovation is out of the scope of the present study object. Organization strategies as regards the hardware contents such as factory modernization, automation, computerization, and robotizing will not be looked into.

0.2. Importance of the study

Japanese-made products, such as TV sets, DVDs, cars, video-cameras, cameras, electronic gadgets, etc. have dominated (still do) the world markets for decades. And that has given rise to the world's sustained interest in the Japanese company. It is generally recognized that the success of the latter lies paradoxically in the reliable quality, relatively low production costs and large diversity of its products. The Japanese company seems thus to prove to the world that, contrary to the conventional wisdom, quality is not necessarily to be associated with high production costs, high salaries and order-made-like production systems.[6] It implies that

---

4   Ch. J. McMillan, *The Japanese industrial system*, Berlin: Walter de Gruyter, 1985, p. 27

5   The software as it appears here contains one more elements than the original by McMillan: the added element is "Production methods". The element is the object of chapter three, whereas chapter one focuses on employment practices, alludes to decision making and information diffusion, but does not deal with social innovation

6   Toyota production system is an order-made mass production system

salaries may be high, the product lines large while at the same time the overall manufacturing costs can, however, be kept low.

From a strict management point of view, I think that in order to understand the strengths of the competitive edger/advantage of the Japanese company, it is important for one to grasp not only its manufacturing techniques but also the management setting in which those techniques are used. Doing so would shed some light on the secret of the competitiveness of the Japanese company.

One should not however lose sight of the fact that a thorough understanding of that matter would include research and development, marketing, finance, legal and other aspects of the Japanese company. This is possible only as a result of a team work. I think however that production and management aspects remain the core of the manufacturing company. And the manufacturing which in Skinner's terms is "the formidable competitive weapon"[7] remains the foundation the strengths of the Japanese company are based on.[8]

0.3. Objectives of the study and work hypothesis

The work hypothesis is that, though JIT looks like a pure production method that is to be associated with industrial engineering,[9] it however takes roots firmly in its birth environment, i.e., the Japanese management. The contention here is that complete understanding of the JIT and its working mechanism would not be possible without, at least, a basic knowledge of its immediate surrounding environment, i.e., the Japanese management.

Concerning the Japanese management, the work hypothesis is that Japanese management should not be viewed as an unstructured set of management characteristics specific to the Japanese company. On the contrary, it will be shown that these features are logically connected like the rings of a chain and make up a consistent system. Furthermore, those management features, besides their internal interconnection, can be grouped into two categories: some may be viewed as management strategies and others as aimed-at objectives. At last, at the deepest level of understanding the Japanese management, it is supposed that there are,

---

7   See W. *Skinner, Manufacturing: the formidable competitive weapon*, NY: John Wiley & Sons, 1985

8   America, with its top business, law and engineering schools, has the best experts in marketing, accounting and finance, commercial law, and research and development, etc. If Japanese products have been very competitive in the world market places, I think it is because Japan has better manufacturing and management methods.

9   See S. Shingo, Study of Toyota Production System from industrial engineering point of view, Tokyo: Japan Management Association, 1981

among these features, some that make up the basic management principles on which lies the structure of the Japanese management system.

In other words, the question of the transferability of Japanese company management practices, from a strict management point of view, becomes a complex one made of many others that can be dealt with at different levels. In fact, a number of questions need to be answered before trying to transfer a Japanese management feature in a different setting. Is a Japanese management feature one intends to introduce in a new environment a strategy or an objective? If it is a strategy, is its objective in line with the company ones? If the feature in question is rather a management principle, can it fit into an environment structure based on different management principles governing the non-Japanese environment?

0.4. Methodology

This study of the Japanese company draws some of its substance from a deep analysis of Japanese management theories and production methods conveyed through books, periodicals, classes, seminars and conferences about the Japanese company.

Besides having the chance to study the Japanese manufacturing company in Japan, I got the privilege of "study-touring" factories and could thus realize how JIT works in Japan. Usually, factory study visits are followed by an exchange of views between researchers and company executives. In some cases, we had the honor to talk to the company responsible number one. Lessons learnt from the factory study tours are also reflected in this book.

Unfortunately, factory visits as well as management theories are limited to big corporations. That is why I initiated a field research. A survey questionnaire was sent by mail to about 440 manufacturing enterprises of Osaka, chosen at random. The survey was done first in October 1991 and a second time in October 1992 through February 1993. In 2003, another survey based on the same questionnaire but of a very limited scale was conducted in Wakayama Prefecture (which belongs to the same Kansai Area as Osaka). A team of 7 people visited each a number of companies, and asked the management to fill the questionnaire. The findings of this limited survey fortunately confirmed however the results and conclusions of the one conducted ten years earlier.

I decided to conduct a survey about Japanese management and production methods for many reasons. First, I wanted to analyze first hand data. Second I was motivated by the desire to know about Japanese management and production methods regardless of the company size.

Besides, I have always felt some uneasiness with theories about Japanese management and production methods. Take for instance the simplest case concerning the lowest level of management rank. For some, it is "kacho" (section head), for others it is "kakaricho" (chief clerk), for some others it is the supervisor. And this is due to the fact that most Japanese management studies are intensive-oriented, limited thus to a very small number of companies.[10] I thought the different and various observations about some Japanese management characteristics might well be accommodated in statistical statements of the type, "In X% of the manufacturing companies, the lowest rank for management is kacho, kakaricho or supervisor".

I had the same feeling of frustration concerning JIT. JIT is said to be used by large corporations only. From time to time, I came however across some people of the small manufacturing who would proudly, convincingly affirm that they run no stocks, they produce different kinds of items on the same line, the die change-over time of their machines is less than 10 minutes. And those are JIT features.

It is generally said that Japanese management and production methods (JIT) concern only big corporations that employ less than 30% of the work force and that they have nothing in common with the small and mid-size manufacturing. If things are so, *how could one explain then the fact that by growing bigger, small companies develop more and more features that are close or specific to those of big corporations*? I thought small companies do not get overnight features they did not at all have before. Management features of the big corporation should be present in

---

10 T. Kagono, I. Nonaka, K. Sakakibara and A. Okumura, (Strategic vs. evolutionary management. A U.S.-Japan comparison of strategy and organization, N.Y: Elsevier Science Publishers B.V., 1985, p.9) complain about the fact that most studies about the Japanese management are only intensive-oriented. In fact, R.T. Pascale and A.G. Athos, in their marvelous book, The art of the Japanese management: application for American executives, NY.: Warner books, Inc. (1982), seem to have considered only one Japanese big corporation, i.e. Matsushita; R. Clark (The Japanese company. Tokyo: Ch. E. Tuttle Co., 1987) refers mainly to one company, Marumaru, a fictive name for a company of which he keeps secret the identity. W. G. Ouchi in Theory Z: how American business can meet the Japanese challenge (NY.: Avon books, 1982 p. ix) said to have done all extensive research: " My research involved hundreds of interviews and thousands of hours of collecting questionnaires and analyzing data. But S. P. Sethi, N. Namiki and C. L.Swanson (The false promise of the Japanese miracle: illusions and realities of the Japanese management system, Boston: Pitman, 1984, p. 266–267) pretend that it is not quite true: " Despite the author's statement that extensive interviews were conducted with a large number of chief executives of American and Japanese companies, both in Japan and in the United States, it subsequently becomes clear that only a handful of interviews were conducted; that all of them were not conducted by the author, that of those interviews, only three were made in Japan and none in the United States. Thus the claim of extensive research would appear to be exaggerated"

the small companies but that they might be at their infant stage or just dormant waiting for favorable conditions to develop and grow.

The quality of the Japanese product is attributed to the effects the Japanese management style has on the work force and to the Japanese production methods. *If Japanese management and JIT concern only less than 30% of the work force and not the more than 70% that produce over 70% of parts of the products manufactured by big corporations,[11] how to explain the fact that the small manufacturing makes those quality parts that contribute to the overall quality of products made by big corporations?* I supposed that JIT should not be completely unknown in the small the manufacturing. And, in order to find a tentative answer to those questions and to verify those hypotheses, I decided to conduct a field research in the manufacturing sector.

The main problem for the survey was that the reply rate was rather low. Many of the 440 companies I had chosen to survey had to be contacted twice or three times before getting their reaction (sometimes. it was a blank questionnaire returned). The effort has somewhat paid off. About 39% of companies got back to us. Unfortunately, of the 169 replies we got, only 129 (about 29% of the 440 contacted companies) were considered to be good or partially good.

0.5. Organization of the book

The book is made of three main parts with two chapters each.

The first part describes management features of the Japanese big corporation and tries to show their internal relationship taking as an observation point one of its pillars, i.e., lifetime employment (chapter one). That theoretical approach is completed by an analysis of data collected thanks to the survey (chapter 2). Due to the overwhelming number of small and mid-size manufacturing companies that characterize the Japanese industry, the survey turned out to be targeting those small and mid-size companies The second part is devoted to the study of JIT (chapter 3) and to the analysis of data of the survey concerning the production methods (chapter 4).

The last part examines the relationship between JIT and the Japanese management system. It tries to integrate Part one and Part two. The integration takes part at different levels.

First JIT techniques are analyzed, and classified. Then their relationship with Japanese management is shown. Second, thanks to the concept of prefiguration the present study is putting forward, the dynamic relationship between management characteristics of big corporations and those of small/medium companies

---

11   See Chusho Kigyo Cho (ed.), *Zu de miru chusho kigyo hakusho*, Tokyo: Okura-sho, 1992

can thus be brought to the light. Third, the prefiguration of JIT is tentatively identified as being in its expansion phase; it seems more appropriate to speak of the emergence of the JIT prefiguration instead of the JIT prefiguration.

Fourth, the same kind of relationship between JIT and Japanese management is viewed also as prefigured in the small/mid-size manufacturing.

At last, due to the close relationship between JIT and Japanese management, especially the strong impact of Japanese management on JIT, the question whether JIT may effectively work in a foreign environment is raised (Chapter 5). The question leads to re-examining and/or rethinking the Japanese management itself. As a matter of fact, Japanese management is here reconsidered and re-understood in terms of its management strategies and basic common principles that underlie its structure (chapter 6).

0.6. Originality

First, the study goes beyond a simple description/statement of Japanese management and shows (1) that the style of Japanese management is a consistent system with elements interlocking into, and interacting with each others; (2) that among what is generally referred to as management features of the Japanese company, there are strategies and aimed-at goals; (3) that the system is underlain by basic management principles whose effects are also part of what is generally called Japanese management features.

Second, not only JIT is described but the nature of its elements is analyzed. The classification of JIT elements sheds the light on JIT elements that can easily be transferred and those that would require a particular environment, i.e., a Japanese-management-like setting.

Third, the study shows the dynamic aspect of the relationship between JIT and Japanese management on the one hand and on the other between big corporations and small/mid-size manufacturing companies through the theory of prefiguration.

0.7. Scope

There are of course some limitations concerning the present study. The study does not go beyond the sphere of management within which it explains JIT. That may seem to be a weakness. But one should not forget that the research focuses on the manufacturing company and that understanding its production method is the main objective. The necessity to deal also with management was felt because of the ramification of JIT elements into Japanese management and the impact of the latter on the former. Due to various constraints of time, it was not possible to go beyond the Japanese management. The limited scope of this study has the advantage of focusing it on its main objective, i.e., the understanding of the JIT system.

0.8. Audience

This book will satisfy the needs of all those interested in the management of the Japanese business enterprises in general and in the JIT/lean production methods in particular. It sheds the light on the extent to which the Japanese lean production methods are also used in the small and mid-size manufacturing companies in Japan. Very often, one will hear that JIT can be applied in big corporations only. This study shows that even small businesses can use efficiently some of its techniques.

People who are interested in intercultural as well as in international management would find satisfaction reading chapters 1, 2, and 6. Those who would like to focus on JIT/lean methods can content themselves with chapters 3 and 4. All those who would like a full understanding of JIT and then examining ways to implement it would learn a lot reading the whole book. At that level, we have in mind the following audience: executives whose companies are competing directly or indirectly against Japanese corporations; consulting firms interested in production methods; manufacturing company managers (regardless of their company scale); management researchers, educators and students. Besides, it is a valued source of information for anyone interested in management in general.

Part I Japanese Management System

Chapter 1
Japanese big corporation management

Chapter 2
Survey-based study of Japanese management in the small and mid-size
manufacturing enterprises

Chapter 1 Japanese Big Corporation Management

The Japanese large scale company has developed an original set of its own management characteristics known as Japanese management. These features have been for decades an object of worship by academics worldwide. The same admiration for the Japanese style of management can be observed indirectly in the high respect imposed by the outstanding performance of the Japanese company.

This chapter outlines first the content of Japanese big corporation management or Japanese management by describing merely and briefly its specific characteristics. Then, it attempts to show that these management traits are logically inter-related, i.e., they form a consistent system.

1.1. Brief description of Japanese management features

1.1.1. Lifetime employment

Lifetime employment means that, as a general rule, a Japanese young man[1] enters a company when he is between 15 and 22 and leaves the same company when he is 60 or older.[2] The first company one starts working for does not offer the first work experience that would increase one's employability by other companies; it is the beginning of the whole experience of the long working career. When a company opens its doors to welcome a new Japanese graduate straight from school, at the same time other companies automatically close for ever their doors to him.

---

1 Women, even those with university degrees are mostly seen in Clark's terms as 'mobile workers, joining the company temporarily and leaving it upon marrying. Although there are more and more career women, their number is still not significant. The Japanese company remains essentially a society of men. That is why the book will refer only to the male worker of the Japanese company. In addition, women do not, as a rule, work on production lines.

2 Lifetime employment as most of the Japanese management features applies mainly to men

1.1.2. Recruitment

Before the long Heisei recession of the 1990s that ended in the mid-2000s, Japanese large corporations used to start the recruitment campaigns during the summer vacation to attract future graduates, i.e., the potential new employees who were to leave school at the end of the academic and fiscal year. By September well before their graduation that usually takes place at the end of March, future graduates knew already what company they would be working for. There was a kind of a respected tacit gentleman agreement between Japanese big companies that prevented them from recruiting future graduates before the summer vacation. During the bubble economy of the 1980s, each university graduate had, on average, more than two job offers. Recruiting at the same time prevented candidates to cumulate too many work promises by different companies. Things have changed since the early 1990s. The long recession resulted in companies recruiting less and less; and in candidates having less job choices. The necessity of the tacit recruitment agreement became obsolete and useless on the one hand; and on the other hand, fourth-year students, i.e., future graduates, started looking for jobs as early as the academic year opens in April. Very few lucky ones might secure firm employment promises as early as May, i.e., almost one year before their graduation that still takes place at the end of March of the following year.[3] Others might wait till January or February.

If one graduates at the end of March without however securing a job at a big business firm, he can be sure he has missed for ever the train to a company of his dream. In fact, the recruitment of new graduates is made once year, in April. There are no other recruitment seasons reserved for former graduates or graduates on the job market!

The recruitment process takes into account the level of education of the would-be new employee, his age, his general aptitude, his family background and above all his attachment to his future employer. At the time of hiring, the candidate's sincere intention of becoming a member of the company he will be working at and "his potential contribution" are "more valued than any other technical skills, knowledge or qualification".[4]

---

3   I am stating this based on my own Japanese experience as a graduate student, a company employee and university professor.

4   Kagono et al. eds., *How Japanese companies work*, Kansai Productivity Center, 1984, p. 140

1.1.2.1. Education

Education is a very important factor in the hiring process. It determines the kind of companies to which you can consider applying for a job as well as it defines the types of companies that can take your application into account. The new employee's starting position and maybe his place of work will be decided according to his education level. In the manufacturing industry, high school graduates are certainly to start working and to spend most of their working life on the production lines. A university graduate even though he starts working as a production worker,[5] his status will sooner than later be changed from the "blue-collar" to the "white-collar".[6] From the time a person is hired, according to the level of his education, he can guess his chances to be promoted to a management post.

More than 50 years ago, Abegglen attracted the attention to the importance of recruitment directly from school:

> *In contrast to the American practice it must be emphasized that recruitment directly from schools into the company is to all intents and purposes the only way in which men enter the firm.*[7]

Since then, things have not changed at all.[8] Table 1–1 shows the evolution of the graduate intake in the manufacturing and the service industries from the mid-fifties to the first half of the 1990s. As one can see it, in 1955, 72% of those who joined the manufacturing sector were middle school graduates; ten years later the figures went down to 55%; and twenty years later only 15 % came from that category; none at all in 1985 and a surprising 6% in 1994. The service industry,

---

5   In Japan, engineers may start working as production workers before being moved to technology and/or research departments. Masanori Moritani, an engineer who began his career on the production line states: "Outstanding college-educated engineers are assigned in large numbers to the production line…Many manufacturing industry executives are engineers by training, and a majority have had extensive first-hand experience on the shop floor" cited by M. lmai. *Kaizen. The key to Japan's competitive success*, Singapore: McGraw-Hill 1991, p. 37

6   In Japan, it is very difficult to distinguish between blue- and white-collars by the work uniform because those who work in a factory wear the same uniform as those who work in offices which is hardly blue! I will however use the word blue-collar to refer to a line worker.

7   J.C. Abegglen, *The Japanese factory: Aspects of its social organization*, Arno Press, 1979 (Reprint of The Free Press 1958 edition), p. 29

8   See the survey results about that topic in Chapter 2

which needs less muscular forces than intellectual ability, has not hired a single middle school graduate since 1975.

High school graduates make up the bulk of these people the manufacturing sector needs while the same category is losing its importance in the service sector. As to the question why the number of high school graduates did not seem so important in the 1950s, it might have been due to the fact that there were less high school graduates than middle school ones at that time. The manufacturing sector was in its high development phase[9] and in need of young people so that it had to get what the labor market could offer then.

Table 1–1 Educational composition intake in selected industries, 1955–1994

Manufacturing graduates	1955	1965	1975	1985	1994
Middle school (15+)	264 (72)	387 (55)	49 (15)	--	16 (6)
High School (18+)	86 (23)	255 (36)	84 (57)	225 (65)	148 (52)
Technical college (20+)	--	0 (00)	4 (01)	30 (09)	4 (1)
College (20+)	2 (08)	9 (01)	20 (06)	81 (23)	27 (9)
University BA (22+)	16 (04)	49 (07)	62 (19)	81 (23)	77 (27)
--{Science, engineering}	--		{34 (11)}	{44 (13)}	{42 (14)}
--{Arts, social science}	--		{28 (11)}	{37 (11)}	{35 (13)}
University MA (24+)	--		4 (01)	8 (2)	14 (13)
University Ph.D. (26+)	--	1 (0)			
Total	368 (100)	701 (100)	323 (100)	384 (100)	286 (100)

Banking, insurance, estate graduates	1955	1965	1975	1985	1994
Middle school (15+)	2 (07)	1 (01)	--	--	--
High School (18+)	21 (72)	65 (82)	69 (62)	28 (37)	12 (18)
College (20+)	1 (03)	3 (04)	15 (13)	21 (28)	22 (34)
University BA (22+)	5 (17)	10 (13)	28 (25)	26 (35)	31 (48)
Total	29 (100)	79 (100)	112 (100)	75 (100)	65 (100)

Source: R. P. Dore and M. Sako, *How the Japanese learn to work,* 1998

The lower the candidate's education level, the less is his chance of entering a big company.[10] The bigger the company, the fiercer is the competition for appli-

9   According to Uemura (*Nihonteki keiei soshiki,* 1992, p. 92), the high development was fuelled by the Korean war

10   I clearly remember the case of a young man that I knew very well, a mechanical high student whose dream was to work for a railway company. When he told me that he was glad because he landed a job in the rail industry, I asked him if he was going to work on the Shinkansen Lines. He replied that as a high school graduate,

cants. It means that the lower the level of schooling or that of the school attended, the less are the chances of passing hiring tests of a big corporation. To make the matter more complicated, in addition to the formal education, companies do take into account the reputation of the university attended. Consequently only the best students who can go to the best schools can easily enter large companies that are considered the best places to both work and spend one's working career in.[11] The relationship between the best universities, the big or best corporations and the best students can tentatively be referred to as a cycle of excellence. Being part of that loop is the dream of any young and ambitious Japanese man.

1.1.2.2. School & Recruitment

Because of the correlation between graduates' knowledge level and that of their schools, some firms seem or tend to recruit only students of specific schools. Abegglen came across a company that hired only from five universities: Tokyo, Kyoto, Hitotsubashi, Keio and Waseda.[12]

I am not sure to what extent the case might be generalized.[13] That case may be linked with the problem of cliques (many old Japanese big corporations have kinds of cliques[14]) due to the fact that the first persons to enter a company tend to recruit from their alma mater. I don't think this might have worked during the bubble economy of the 1980s during which period there were more job offers than job seekers (graduates).

What may be quite true is that, all other things being equal, if a Japanese company has to choose between two candidates, the one from a name university would be given top priority over his competitor from a less reputed university.[15] Anyway, the thing worth bringing to the light is that the type and size of a company one

---

he can't apply for a position at the Japan Rail Company. He just was going to work for a regional railway company, the Nankai Line. On the other hand, Kiyo Bank, a regional bank, no more hires high school graduates.

11 Big corporations pay higher salaries, offer more job security and are thus sought for by many applicants. In order to get a better job, it is necessary not only to go to college but to enter a good university. And the competition for a university recognized as good is very fierce.

12 J. C. Abegglen. *The Japanese Factory. Aspects of its social organization.* 1979, p.30

13 The 20 years I have spent in and outside Japan studying Japanese management and production methods, working for a company and teaching at the university level can not help confirm whether this is a general or a particular case.

14 R. Clark, *The Japanese company.* Tokyo: Ch. E. Tuttle Co.. 1987, p. 163

15 This is not something specific to the Japanese company only

may enter depend much more on the level of education. And the school attended is a major factor affecting the education level as perceived by the public. Clark summarizes the importance of education as regards the recruitment process:

> *A good education is rightly seen as a passport to a big company with its prospects of high pay and greater security. A middle school leaver will have less chance of working for a major engineering company than a high school leaver. A man from a poor university will be unable to join a major bank. The education can be seen as a pre-selecting mechanism for the labor market.*[16]

1.1.2.3. Skills

As for high school graduates, the type of studies done seem not relevant for most companies as shown in Table 1–1.

It seems clear that the majority of firms of both the manufacturing (59.25%) and the service (63.1%) have most of their employees recruited neither in the technical nor in the commercial schools, respectively.

It is important pointing to two things. First, the manufacturing is not, in general, inclined to recruit women for the posts of line operators. Second, the selection of women in the service sector seems to emphasize the type of education the new female employees have had. Accordingly, the majority of firms (53%) in the service industry hired female commercial high school graduates. In spite of that, a very large minority (47%) recruited female non-commercial high school graduates.

Concerning the fact that companies seem to have different education criteria in hiring women or men, the explanation may reside in the fact that women are not (expected) to be working lifelong for a particular company. Japanese working conditions for regular employees are such that, in general, it is preferable for women to leave the company as soon as they get married.[17] For such category of employees referred to by Clark as "mobile workers", it seems reasonable to hire those with the needed special knowledge since it does not make much sense to spend company resources training them as they will not remain with the company for a long period of time. The case of women is like that of a foreign university graduate in Japan: he is sought after mainly for his or her special knowledge.

---

16   R. Clark, op. cit. p. 146

17   Long working hours, drinking with colleagues and superiors after work (nominication) result in getting back home very late for a Japanese "salaryperson". In such conditions, it is difficult to cumulate housework and professional job. See also Clark, op.cit., pp. 143–145

The majority of firms that hired high school students with special knowledge and skills related to the manufacturing or service sectors stated however that if they had to hire ordinary high school graduates, they would do so because "they may be trained in-house to do various types of jobs". That reason is mentioned by 86.4% and 65.8% of the companies of the manufacturing and service industries respectively. As for women, only a minority (46.6%) said they would do so for the same reason. This means that only a minority of companies would accept or feel ready to train women.

In the hiring process of university graduates, the Japanese company does not stress a special skill or knowledge. The potential to acquire and develop specialized skills at the work place is the point much paid attention to. New employees may be roughly divided into two main categories: engineering & science and social science & humanities.

1.1.2.4. General aptitude tests & interviews

As Japanese corporations do not emphasize special training or knowledge when hiring new employees, they require applicants to be potentially capable of performing many kinds of jobs. Thus the recruitment in Japan is mostly based on aptitude tests[18] and on many levels of interviews aimed at detecting the potentials of new graduates.

1.1.2.5. Age

Japanese big corporations recruit in what Clark refers to as the primary labor market[19] which is made up of young graduates fresh from school without any prior work experience. In other words, they let in only young people whose age varies between 15 and 22. Those are middle school, high school, junior college and university graduates. The importance of age can be seen from the fact that graduating from the university at age 25 or 26, whereas the average is 22, consitutes a serious disadvantage and a handicap for such candidates during the hiring process. As a matter of fact, age is paid attention to during the hiring process. One has to remember that chronological age is taken into consideration in the calculation of salary and its annual incremental increase. All other things being equal, younger new employees are cheaper than older ones. That might be why they are given priority. It is just economics.

A bright graduate Brazilian of the Japanese descendant had problems getting a job because he was, according to his own accounts, older than the average. At

---

18   J. C. Abegglen, *The Japanese factory*, p. 32

19   R. Clark, *The Japanese company*, pp. 141- 149

the university I worked, when there was a need for new teachers, the age range was always specified during the faculty meeting although the published vacancy announcement would mention only the job title (full, associate or assistant professor), the field of teaching and research interest (accounting, production, etc) without any allusion to the candidate's age.[20]

1.1.2.6. Family background

The Japanese company requires records from the applicant's family and personal history. It implies that because of the applicant's social origin, he may be denied the job even though he has successfully passed all the tests. By the way, Ouchi who seems to have studied in depth Japanese and Japanese-like companies to which he coined the name of "Z companies" qualified them to be sexist and racist.[21]

In Japan where you have only one race the Japanese company in Japan can be said to be sexist and discriminatory, rather than racist. The discrimination is based on the family's social status.[22]

One of the points the Japanese company seems to attach so much importance to is the fact that it wants to be seen by the potential employees as their new family. Therefore, the biological family of each candidate is considered like a mirror reflecting the future behavior of the potential employees. Conversely the importance of this point can be shown in the fact that a family member could see the conditions and the premises under which his relative is going to work.[23]

The hiring is like a marriage. People have to make sure the partner is from a good family. The company expects the new member not to leave it as well as the new member considers the company not to let him go.

1.1.3. Mutual commitment

The Japanese work force is referred to as representing a fixed cost. In other words, the staff volume does not swell so easily because the company is doing well and then burst so easily during hard times. Hiring is like a Christian marriage where the two partners promise to live together for the better and for the worse.

---

20   There are many unwritten rules in Japan. Of course, no one is told that he is not given a job because of his age. It is against the Japanese law

21   W.G. Ouchi, Theory Z: how American business can meet the Japanese challenge. N.Y.:Avon books, 1982, p. 77

22   There are consistent and founded rumors that a social class of Japanese people called Buraku is discriminated against in the work force market.

23   R. Clark, *The Japanese company*, p. 1 58

Concerning this point, an experience of Sony Corporation may be recalled because it is very informative.

In 1955, Sony made its first small and simple transistor radio. Then Mr. Morita took it to the USA to look for a possible market. With a production capacity of less than ten thousand radios a month, an American retailer proposed Morita a deal of one hundred thousand radios. It seemed to be a golden opportunity for a young company in search of a source of income and of the market of its new product. The reaction of Mr. Morita was surprisingly contradictory.

> *Our capacity was less than ten thousand radios a month. If we got an order of one hundred thousand, we would have to hire and train new employees (…)..If we would not get a repeat order the following year we would be in big trouble (…) because how could we employ all the added staff (…)? In Japan, we can not just hire people and fire them whenever our orders go up or down. We have a long-term commitment to our employees and they have a commitment to us.[24]*

I am afraid the rationale that led to rejecting the offer would sound acceptable only in the Japanese business context.[25] In other countries, managers' performances, especially financial ones, are monitored closely. And their year-end bonuses are conditioned by those performances. In such a context, it would have been really a golden opportunity to seize absolutely.

But the Japanese management environment is different. When a Japanese company lets someone in upon his graduation from school, it will keep him in his working life long. Interesting, a graduate who starts working defines himself and is referred to as "shakaijin", a member of the society or community. And a community doesn't expel or banish its members except for criminal acts. And being expelled is a harsh punishment:

> *Once hired, the new employee is retained until mandatory retirement at age fifty-five. An employee will not be terminated for anything less than a major criminal offense and termination is a harsh punishment, since the one who has been fired has no hope of finding employment in a comparable firm and*

---

24 A. Morita, with E. M. Reingold and M Shimomura, *Made in Japan Akio Morita and Sony*, Glasglow: Fontana/Collins 1987, pp. 85-.86

25 Almost all the Japanese to which I mentioned the reaction of Mr. Morita thought he could have accepted the offer. One thing is sure, at his place, they would have seized that opportunity. Anyway, nobody will never know whether he took a right decision or not. If the retailer was a Japanese, he would not have hesitated to accept the deal.

instead must turn either to minor firm that pays comparatively low wages and offer little security, or else must turn to his hometown.[26]

Reciprocally, the same attitude is expected from the new employee. The company that accepts him as a new member expects him to remain a permanent member. He is not expected to change companies for the sake of his own selfish interest or happiness, for a better salary or a better working life elsewhere. He should do his best to make his work place the better place to work at and live in.

The Japanese company, as a rule, does not welcome deserters because they are regarded as traitors. And that attitude prevents people from moving from a community of work to another.

A Japanese young person becomes a "shakaijin", i.e., a member of the society when s/he starts working, i.e., when s/he serves the society. The company is the organization through which one serves the society. If someone has to change communities s/he might be accepted but as a "neophyte", needing to be re-initiated to the new community. The initiator may be chronologically speaking younger than the new member even though the latter was a senior in his former community.[27]

1.1.4. Leaving the Japanese company

There are three ways of leaving a company: retirement, lay-off, and resignation.

1.1.4.1. Retirement

In Japan, retirement is the common way of leaving a company. The Japanese worker commits to his company and the latter commits itself to the former so that there seems only one natural, or normal way of leaving the Japanese company, i.e., by retirement.

The notion of retirement in the Japanese business context needs however some clarification. You have the mandatory retirement and the voluntary retirement. The mandatory retirement age is the age at which a company employee ceases his activities as a regular full-time worker (he may, however be reconverted into a part-time worker). The retirement age is fixed by internal regulations of companies.

---

26   W.G. Ouchi, *Theory' Z: how American business can meet the Japanese challenge*, NY: Avon books, p. 15

27   I have in mind the case of someone who defined himself as a bucho student or a bucho in training. He had retired from Sumitomo Heavy Industry and landed a new job in the real state business as bucho (division or department head). He was being initiated by younger colleagues who were lower in rank!

The voluntary retirement is full of subtleties; it may mean voluntary retirement in the true understanding of the word, i.e., someone wants for a reason or another to quit the company earlier than the mandatory retirement age. It also refers to (1) forced retirement by the company upon a worker; (2) early retirement sought by and encouraged by the company for some reasons, and especially for economic or financial ones. In addition, when division or department managers (heads) are promoted to the board of directors, they cease, according to the Japanese business law, being company employees, i.e., they have to retire in order to become employers' representatives. This is a legal and theoretical retirement.[28] I have never met a Japanese high executive who regards himself as having retired his company because he has become a member of its board. Acceding to the board of directors is rather viewed as an important promotion.[29] It means also that you will retire much later than those with whom you started working together for the company. The higher you go up into the organization hierarchy, the longer you'll be working for the company.[30] The only one who reaches the top position of president will never retire from the company.[31]

1.1.4.2. Resignation and lay-off

Resignation means terminating on his own will one's contract with the company and leaving that company whereas lay-off means being let go by the firm.

In Japan, laying-off employees tarnishes the company image from both inside and outside. As layoff policy or practices may lead to not attracting any more young bright graduates from good universities and demoralize the company workforce, it is often disguised as resignation. For the good of the company as a community, the individual may thus be sacrificed and his reputation may be stained: he may look like someone who cannot commit himself to his working community, i.e., the company he used to belong to.

---

28 That kind of resignation or retirement is just a legal formality. At the company I worked for, a few people were promoted to the board without any sign of their resigning from one position to take another. It was just a promotion.

29 The case of T. Ohno's ejection from the board is very instructive and the effect it had on factory employees such as Kinoshita who left Toyota to set up a consulting firm (For details, see l. Shinohara, NPS no kiseki be-ru o nuida seisan hoshiki, Tokyo:. Toyo Kezai Shimbun-sha, 1985)

30 R. T. Pascale & A. G. Athos. *The art of the Japanese managemen'. application for American executive*s, N.Y: Warner books, Inc. 1982, pp. 246–247

31 Former company presidents end their career as top advisors for life.

The disguised lay-off is referred to as "yamete morau", i.e., accepting the employee's resignation instead of "yamesaseru"[32] which means making someone resign or forcing him to do so, i.e. to dismiss him. In the "yamete morau" case, the image and honor of the individual might not always be negatively and badly affected. As a matter of fact, in Japan, there are many culturally and socially accepted, even respected, sound resignation reasons. Quitting one's job in order to care for the aging parents or to take over and run the family business for example is commonly well regarded. Being laid off may mean just being banished, i.e., being terminated for a major offense and that does not sound honorable.

1.1.5. Promotion

The promotion is made essentially and almost exclusively from within. If a company recruits experienced workers or managers, it usually means that it is a new company or the person is a talent as testified by Morita in the case of Sony:

> *When Sony was new and small, we could steal people from other companies and get away with it, but now we are so large it is not considered the right thing to do, although we still keep on scouting for talent.*[33]

The promotion depends on the following elements: age, length of service, merit, and education.

1.1.5.1. Age

The promotion in the Japanese big corporation takes the factor 'age' into account. The Japanese company looks like its political counterpart where at the top you have only very old people. The prevailing spirit seems to be: "Everyone has to wait for his turn".

---

32 I know a small travel agency that had a staff of seven people, five men and two young ladies. After some time, I realized a permanent absence of one lady and I was told she had resigned. A week later, I met her in a department store where she had started working as a shop clerk. She told me that they let her go. Back to the company, I raised the issue. Then they replied they had to let her go for the survival of the company.

33 A. Morita, With E. 54. Reinhold and M. Shimomura, *Made in Japan*,1987, p. 160

1.1.5.2. Length of service

The length of service is correlated to the age since people are hired from school. In Japan, one can tell someone's age only by asking when he/she finished his/her primary, secondary or university education.

As employees are hired once a year at their graduation, new comers are younger than those with one or more years of service or experience. One may rightly wonder whether the length of service should be emphasized over the chronological age. That view point may be supported by the fact that changing jobs implies almost always restarting the working life at a lower level regardless of one's age.

However, when people are sought after because of their special talent, they usually start with a higher rank and may even get a fast promotion although they might not have yet served the new company for a long period.[34] The case of Ohga that Morita convinced to join the Sony Corporation is very eloquent:

> *Ohga finally joined the company as general manager of professional products and in a year and a half was in charge of all consumer tape recorder operations. In 1964, after only five years in the company, and when he was only thirty-four, he became a member of the board, something unheard of in the traditional Japanese company.[35]*

1.1.5.3. Merit

The promotion in the Japanese company is based not only on the age/length of service but also on merits. It means that capable persons who have accumulated the required experience can continue to climb the management ladder. At the starting step, the promotion is almost automatic for everyone but as there are fewer and fewer posts above, the competition tends to become fiercer and the merit as a factor weighs more heavily.[36]

Matsushita put it this way:

---

34 This is similar to what takes place in other countries while changing companies. Here in Japan, that sounds unusual if not exceptional. But, the development of new areas such systems engineering and other computer-related new fields, that seems however to become the rule even in Japan

35 A. Morita et al. Op. cit., p. 160

36 There is also what should be termed political factors that influence the promotion to the top but that consideration is out of the scope of that study

Another part of the policy was that we would continue to do what we had done before, to select for certain jobs from among our employees who were competent and diligent, and who had performed meritorious services.[37]

In exceptional situations, merits and ability might be given precedence over length of service and age as illustrated by the nomination, in 1977, of one of the youngest board members to assume the function of the president of Matsushita Company:

My son-in-law, Masaru Matsushita, became board chairman in 1977; Toshihiko Yamashita succeeded him as president. The selection of Yamashita for this key position caused quite a stir, for he was one of Matsushita's youngest directors at the time. But despite his youth, we judged him to be the man most capable of guiding the company in its efforts to respond to the rapid social changes that were taking place.[38]

The case of Ohga of Sony Corporation and that of Yamashita of Matsushita Company constitute convincing arguments regarding merits and ability as a decisive factor for promotion, especially at the corporate top management level.

1.1.5.4. Education

It is an undeniable fact that, in the Japanese context, any employee, the production worker as well the white-collar, can become a manager. It is however worth noting that the level of education plays an important role. Not so many non-university graduates go higher than the rank of section head by the time they reach the mandatory retirement age. On the other hand, college graduates have more chance of becoming department or division heads.

This is not a strict restriction. A bright and capable high school graduate can go up so high as to reach the rank of a director or the company president. Taiichi Ohno, the father of JIT production, who had graduated from a mechanical technical high school,[39] made it to the rank of vice-president of Toyota Corporation.[40] Another case worth mentioning is that of Tetsujiro Nakao of Matsushita Electric

---

37 K. Matsushita, *Quest for prosperity: the life of a Japanese industrialist.* Kyoto: PHP 1988, p. 168–169

38 Ibid. p.329

39 This kind of technical high schools of the pre-war Japan correspond in fact to the present junior technical college

40 See T. Ohno, *Toyota Production System. Beyond large-scale production.* Cambridge, MA.: Productivity Press, 1988,. especially the notice about the author

Company. Though without any formal education, he climbed, thanks to his merits, the management rank ladder and became one of the highest officials in charge of the technical department.[41]

1.1.6. Salary

The salary in Japanese corporations depends, like the promotion, on education, age, length of service. To those elements one should add the scale of the company.

1.1.6.1. Education factor

Formal education affects mainly and directly the starting salary. The middle school graduate is fifteen, the high school graduate is eighteen, the junior or technical college leaver is 20 and the university graduate is 22 years old. Education affects directly and determines the level of basic salary.[42]

Table 1–2 Lifetime earning by educational level and size of firm, 1985

University graduates (Overall=100)				
Enterprises employing	Base Pay	Bonus	Pensions	Total
>1000	100	100	100	100
300-999	88	80	80	85
140-299	84	74	68	80
<100	82	67	66	77
High School graduates: Clerical & Adm. (overall = 95)				
>1000	100	100	100	100
300-999	94	86	96	91
100-299	89	78	79	85
85 <100	89	67	70	81
High School graduates: Production workers (Overall = 82)				
>1000	100	100	100	100
300-999	101	98	108	101
100-200	91	81	80	87
<100	89	67	70	81

Source: R.P. Dore and Sako, How the Japanese learn to work, 1989 p. 31

41　K. Matsushita, *Quest for prosperity*, pp. 106- 113, 135–38, 176–177 and 265

42　Japanese companies job offer in their internet sites specify the basic salary by level of education

1.1.6.2. Company scale

As stated just above, the starting salary is a function of the level of education and of the enterprise size. Top companies pay higher wages while smaller companies pay lower salaries. Tables 1–2 &1–3 give the differential rate of life-time earnings and wages for big, mid-size and small companies. The pay is determined at least the starting one, by the level of education, as stated earlier.

Table 1–3 Wage differentials by size of establishment

Establishments emplying	1970	1980	1985
More than 500 workers	100.0	100.0	100.0
100–499 workers	81.5	80.5	77.1
30–99 workers	81.6	65.3	62.9
5–29 workers	61.8	58.0	54.9

Source: R. P. Dore and M. Sako, How the Japanese learn to work, 1989, p. 31

1.1.6.3 Age & length of service

Age and length of service contribute to the yearly increase of the Japanese salary. They are given different weights; but the weight is proportional to someone's age and length of service. That is why people of the same age who entered the company at the same time will get the same basic salary provided that they have the same level of education.

As the chronological age of young people corresponds, in Japan, to the level of education, how can one know the individual impact of age and education on salary? The truth is that age and level of education of the new employee are disassociated and considered separately. The age (as well as the length service) related allowances or rewards are available grids of salary and rewards in the personnel department.[43]

In Japan, even the regular salesman is not paid on commissions and bonuses. His salary is based as for everyone else on his age, length of service and education. Huddleston, Jr., gives the following management advice to any foreigner wanting to run a business enterprise in Japan.

The salaries of sales personnel should be the same as for all other personnel of the same age, education background and experience.[44]

---

43 The age's monthly allowance for a small company that provided that information during my survey study is as follows: ages 16; 18;, 20; 22; 26; 30; 40; 50; attached rewards are 13000; 14000; 15000; 16000; 18000; 20000; 25000; 30000 Yen respectively

44 J. N. Huddleston, *Gaijin kaisha. Running a foreign business in Japan*, Tokyo: Ch. E.Tuttle Co., 1990, p.205

The basic salary is increased every year for everyone. The basic salary as well as the annual incremental increase takes into account the chronological age and the length of service. If it happens for someone of a certain age and experience to change companies, he is likely to get a starting salary that will be lower than that of other workers of the same age and education but who has joined the company years earlier. He would get less pay since his length of service at the new company is zero and his past experience (length of service) at the former company might not be 100% taken into account.[45] The age allowance will be however similar to other workers of the same age.

1.1.6.4. Allowances and bonus

The total earning of a Japanese company worker is made up not only of the basic salary but also of a bonus and many other kinds of allowances. Those allowances are not always related to the work and company performance, a fact remarked by Abegglen many decades ago. One of these special rewards is the bonus that is usually paid twice a year, in June and December. In some companies it is equivalent to six times or more the monthly salary. At least it is worth two months of salary.[46]

1.1.7. Company labor unions

The worker of the Japanese company hardly belongs to a supra- company union (national or regional) because of his skills, qualification or the nature of work he is doing. As a general rule, each company has its own labor union[47] to which belong all its unionized employees, regardless of the diversity of their functions or the nature of work/operations they are performing. Such kinds of unions are referred to as company unions.

By the way, in Japan where there is no clear distinction between the scope of someone's job and that of the fellow worker next to him, it is almost impossible to imagine a trade union based on specific skills and/or on the nature of work.

---

45 See Chapter two

46 For more details on this matter, refer to Chapter two and to L. Kupanhy, Japanese management in the small and mid-size manufacturing: a survey", *Keiei Kenkyu*. Vol.43. No.3, 1992, pp. 47–61

47 Company labor unions are members of national federations of unions. A national federation does not deal directly with individual companies but it issues directives individual company labor unions may follow during the negotiations with their company management

The Japanese worker is mainly multi-skilled, generalist, capable of performing a broad range of jobs simultaneously. Interestingly, it is the task of the company to help him develop and acquire a multitude of skills that the company needs. That is why, concerning the unionization, it seems better in Japan to speak of the labor union instead of trade union.[48]

In France, like in Japan, labor union membership does not depend on special skill or the nature of work one may be performing. The main difference is that French labor unions are national organizations whereas Japanese ones are in-company organizations.

1.1.8. Managers

It is interesting to note that the personal history of those holding management posts varies greatly: there are former labor union members, recruited managers and retired officials from the Japanese government ministries.

1.1.8.1. Labor union members

"In many of large companies, the labor agreement stipulates that when an employee is hired he will simultaneously become a member of the union within the company".[49] The majority of Japanese managers having been promoted from within were union members; some of them were even union leaders (see Table 1–4).

Table 1–4 Proportion of directors who have experienced the executive committee of labor unions

Number of companies with directors who have been an executive committee member in labor union	232 companies	74.1%
Number fo companies answering	313 companies	
Number of directors who have been an executive committee member in labor union	992 persons	16.2%
Number of directors answering	6121 persons	

Source: Kagono et al. *op. cit.* 171

---

48   Although this is out of the scope of this study, I think however that only the union by skills or trade deserves the denomination of trade union.

49   Kagono, op. cit. p. 169

High school graduates

Line workers are usually high school graduates. After some years working on the production lines, some of them will become team or group leaders. Usually, labor union leaders or representatives are drawn among those people. Once promoted from team/group leader, the worker ceases being a member of the labor union and becomes a manager. The rank just above that of team/group leader is that of deputy section chief. Upon reaching that rank, one crosses the border and becomes a member of management.[50]

University graduates

University graduates start working in the personnel, sales, research, engineering or R&D departments. Their starting rank is the lowest one and is the same as that of high school graduates: they are just "hira-shain" or ordinary company workers (this is a rank that is never printed on the meishi or business card). In their capacity of hira-shain, depending on the company union regulations, most of them are automatically union members. They may even begin their working career as line workers in the production department for some time. They know, however, as does the company that they will for sure end up, sooner than later, holding management posts in the future after they have built up their seniority and proven their abilities.

1.1.8.2. Recruited management

Unless the company is at its early stage, it will not recruit managers from outside. However, it happens from time to time that companies try to recruit their competitor's managers but this is rather something very rare in Japan. Recruited managers are not job hunters. They are hunted and sought after by companies because of their talent, their particular technical knowledge and skills.[51]

1.1.8.3. Retired government officials

Former government officials, especially those from the former powerful Ministry of International Trade and Industry,[52] or the Ministry of Finance are sought after

---

50   See Clark, R., *The Japanese company* 1987; A. Morita, *Made in Japan* 1987; Nihon Keizai Shimbu-sha (ed.), *Terase de yomu nihon no keiei*, Tokyo: Nihon Keizai Shimbun-sha, 1959

51   For the time being, system engineers and Software engineers are the most targeted by companies

52   It is now known as Ministry of Economy, Trade and Industry (METI)

by corporations. Retired government officials are usually recruited as members of the board of directors. The advice of Huddleston, Jr., a manager and a consultant for foreign companies in Japan is very instructive on that point:

> *Finding the right Japanese chairman is a critical recruitment process in itself, however. The Japanese government is a good place to start, but this is also where Japanese corporations themselves usually start when they have a management gap to fill. A natural choice is a candidate in the ministry that has responsibility for the industry in which the business operates.[53]*

Those former government officials facilitate the company contacts with the government.

1.1.9. Labor union and capital

The life and survival of any company depend, among others, on the relationship between the capital, the management and the labor or workforce. Management occupies the central position since they interact and deal directly with both shareholders (capital) and the labor.

1.1.9.1. Management & Shareholders

The independence of Japanese management vis-à-vis shareholders is a well known trait of the Japanese company. Managers feel so free that in many cases the outgoing chief executive officer designates his successor.

> *According to a survey by Nikkei Business Review, 69.6% of presidents said they decide who will succeed them.[54]*

Japanese managers are not people brought from outside by shareholders in order to look after the interest of the latter, i.e., a quick return on their investment. Because of their independence, the dividends paid to shareholders is said to be very low in comparison with the American companies:

> *Another factor that may have limited the response to competition in the copier market is the high dividends paid out by Xerox and other American corporations in comparison with Japanese firms. Xerox paid $300 million in dividends for 1986, nearly 65 percent of that year's net earnings. A com-*

---

53   J. N. Huddlestonlr, *Gaijin kaisha*, p. 31)

54   Nihon Keizai Shimbun sha (ed.), *Terase de yomu nikon no keiei*, p. 116

parable example is IBM, which paid 43 percent of its 1985 earnings as shareholder dividends. Big Japanese companies pay little or no dividends.[55]

Management people feeling independent from shareholders, their agreement is necessary before a takeover or an acquisition of the company by a third party can be achieved. Free of pressure from the capital, Japanese managers care much more for the company interest and for those who work for it.[56]

1.1.9.2. Management and labor unions

The labor/management relation in the Japanese company is not of the bargaining, opposition, or of antagonistic type. Japanese management and labor union develop smooth relations based on collaboration, cooperation and mutual understanding.[57] Both management and labor union participate in the company growth.

By the way, it is known that Japanese managers care much more for the workers from which some of them came and among which some others will emerge as tomorrow's managers. In general, employees respect managers because they know that tomorrow it may be their turn to become managers. Managers also respect workers because they were themselves, yesterdays, ordinary workers of the same company.

1.1.10. Job rotation

The Japanese worker does not regard himself as a chief engineer, a senior engineer, or an account leasing or selling his expertise and/or technical knowledge to the company. He thinks of himself as a member of the company community. He is serving in such a capacity for the time being. He spends his entire working life in the same company doing many kinds of jobs, moving from department to department, from region to region. This is true even for the line worker though his mobility is confined mostly within a plant. A Japanese production worker can not define himself just as a driller, for example. He is rotated from group to group, from section to section, from a kind of machine to another, from a process to a different process.

---

55 See M. L. Detouzos, R. k. Lester and R. M. Solow, *Made in America. Regaining the competitive edge*, N. Y.: HerperPerennial, 1990, p. 274

56 See J. C. Abegglen & G. Stalk, Jr., Kaisha, the Japanese corporation, Tokyo: Tuttle Co., pp. 181–213

57 K. Urabe. "Innovation and the Japanese management system" in K. Urabe, .J. Child and T. Kagono (eds.), *Innovation and management. International comparison*, Berlin: W. de Gruyter, 1988, p. 12

1.1.11. OJT

On-the-job training (OJT) is a training undergone at the work place in order to learn new techniques, new methods, new work routines or how to handle machines and equipment one has never operated before. As a general rule, OJT is the task of the senior worker or the team leader upon whom falls the responsibility of training young workers or a group of operators by initiating them to the job.

In a word, it is the task that consists, for a knowledgeable person, to teach and transmit his know-how to other persons recently assigned to a specific job he masters well.[58]

1.1.12. Off-JT

Off-the-job training (Off-JT) is not undertaken at the work place: it is not a hands-on-the-machine work training. Off- JT takes place in a company's training center; it may also consist of evening classes or lessons by correspondence.[59] The Japanese company usually lets learning workers support the cost of their studies and reimburses later successful ones.[60]

1.1.13. Group management

Scholars have tried to term Japanese management as group-oriented management or group-centered management.[61] Let us, as Morita did, call it simply group management.[62] Group management means management by the group for the interest of the group.

The concept of group is very crucial to understanding the notion of Japanese management. Japanese people work by group and a group constitutes the basis of cooperation and competition within each company. Members of the same group cooperate closely and different groups emulate and at the same time they cooperate.

---

58 See H. Terasawa, *OJT no jissai*, Tokyo. Nihon Keizai Shimbun-sha, 1989

59 Ibid. p. 12; Dore, R. P. and M. Sako, *How the Japanese learn to work*, London: Routledge, 1989, pp. 82–88

60 Employees who are sent by their companies to study abroad usually do so at the company expenses

61 See H. Hazama, *Nihonteki keiei: shudan shugi no kozai*. Tokyo: Keizai Shimbun-sha, 1971, Odaka,K., Japanese management: a forward-looking analysis, Tokyo: Asian Productivity Organization, 1986

62 A. Morita, *Made in Japan*, p. 199

The effectiveness of organization activities is raised by competition among groups. Cohesiveness among intra-group persons means in itself confrontation with other groups ... Groups are in competitive relationship. But on the other hand, the competitive relationship changes to cooperative relationship when each group is taken to be a subgroup of the larger group.[63]

1.1.14. Decision process

The decision process in the Japanese company is one of the least clear elements. It is difficult to grasp though its effects are very tangible. It is said to be a matter of consensus during a meeting. But it is rather difficult, especially for a foreigner, to understand whether and/or when it was agreed upon.

The idea for a decision may originate at a lower level. The Japanese company looks sometimes like the United State of America where laws projects are initiated either by the president or by a congressman. Then, bills are discussed, debated, amended, accepted or rejected in the Congress. The president also may accept or refuse to sign into laws Congress propositions of laws. In any case, agreed-upon bills by both parties bear the signature of the president and take corps as laws. In the Japanese company the proposition for a decision may start at a lower or higher level of management. It is going to be considered by everyone concerned, discussed and if the consensus is reached from bottom to top and or from top to bottom, the high management acts it as a decision to be implemented. Morita comments on the notion of the consensus for decision:

Gaining consensus in Japan often means spending time preparing the ground work for it, and very often the consensus is formed from the top down, not from the bottom up as same observers of Japan have written. While an idea may arise from middle management, for example, top management may accept it whole or revise it and seek approval and cooperation all down the line.[64]

The consensus of the Japanese decision-making should be understood as follows:

Once a decision is reached whether it originally came up from the shop floor or down from the front office, it is the Japanese way for everyone to devote

---

63 Sh. Uemura, "The Japanese way of management: its characteristics, current practices, and future perspectives", *Osaka City University Business Review*. No. 2, 1989, p. 21

64 A. Morita, *Made in Japan*, p. 198

every effort to implementing it without the sniping and backbiting and obstructionism that is sometimes seen in some Western companies.[65]

The consensus consists mainly in accepting to do enthusiastically the option sanctioned by the top executive.

The most important point here is the fact that the Japanese decision process involves more people than it would be thought of in an American company where only the top management acts like the brain of the company and possesses the power, the right and the duty to suggest and make decisions.

1.1.15. Japanese top executives

Japanese executives are usually raised in the company. Such executives are impregnated with the company philosophy and culture. They are people who have risen to their high positions not only because of their seniority but also because of their abilities and leadership qualities. Decisions they take are in accordance with the company philosophy and are not conflicting with the highest interest of the company and its community.

The top executive is a grown-up child of the company he has become in charge of. He is like the eldest son of the family who takes over the family business from his father. By the time one reaches the rank of president, usually all those who entered the company with him at the same time have already retired. He looks thus like a respected family head.

1.2. Japanese management as a consistent system

The synoptic description above is based on academic approaches to the concept of Japanese management. It might have left the impression that the different management characteristics examined so far are sporadic and un-related to each others. That would be an error of perception. In fact, we contend that most Japanese management features are so closely related, so logically intertwined into each others that they form a consistent system.

There are some key features around which some others are clustered. It is often said that the Japanese management system revolves around what is referred to as Japanese management pillars.[66] They are lifetime employment, consensual decision making process, seniority-based pay and promotion system to which one

---

65 Ibid., p.199

66 Ch. B. Hilton, "Japanese management: clockwork or chrysanthemum at American perspective", *Osaka City Business Review*, No.3, 1991- 1992, pp.67–73) distinguishes three management pillars, i.e.. lifetime employment, consensus decision making,

may should add labor/management & management/shareholders (capital) relationships. Each pillar can be considered a smaller system within which other features or elements are interconnected and the pillars themselves are interlocked. This implies that the whole structure of the Japanese management system can be grasped from a pillar taken as a point of observation.[67] This second section of the first chapter consists in an attempt to show the consistency of Japanese management as a system essentially from the perspective of lifetime employment.

1.2.1. Lifetime employment

The Japanese system of career-long employment which applies only to men has several key elements. First, the employee is hired directly from school, rather than from an open job market. Second, he is hired for general characteristics and abilities, rather than for a particular skill or a particular job. Third, he is expected to remain with the company for a life-long career, and in turn expects not to be laid-off or discharged.[68]

Lifetime employment presumes that the company commits itself to the worker and the latter commits himself to the former. There is a necessity of mutual commitment. The process of lifetime employment consists of the logical interconnection of the following elements: recruitment, mutual commitment, training, promotion & reward system, and mandatory retirement.

1.2.1.1. Recruitment and initial training.

Hiring only new employees fresh from school means letting in quite inexperienced people who don't know how to accomplish the company work. They are not yet able to perform the value-creating and related supporting operations that are necessary for the life and survival of any enterprise. They have neither the knowledge nor the skills required for such operations.

One should note that the Japanese university system does not, as a general rule, educate students for a specific type of jobs. On the other hand, the Japanese company does not at all expect the university system to give students such a specific knowledge required at the work place. Hiring graduates straight from school implies therefore another feature, i.e., the necessity for the company to train the recruits before assigning them specific jobs or tasks to perform (Figure 1–1).

---

seniority-based pay/promotion. I think it is necessary to add the labor management relation.

67   This is what I will try to show

68   J. C. Abegglen & G. Stalk, Jr., *Kaisha. the Japanese corporation* 1987, p. 199

Figure 1–1 Lifetime employment and training

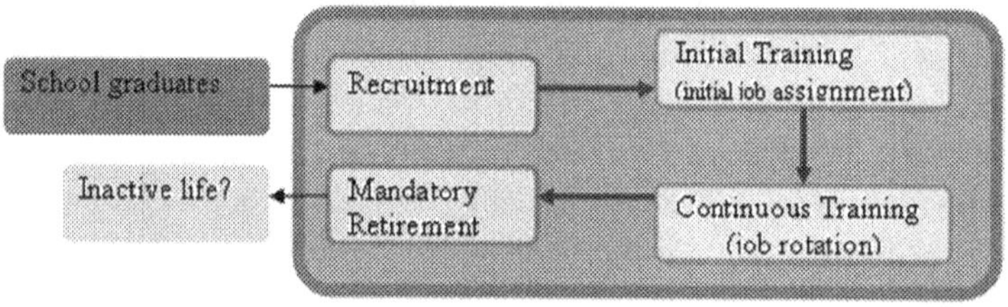

Training within the company means giving new employees the necessary skills the company needs. In other words, the company has to trust new workers upon hiring them and the latter should trust the former. The complicated hiring process that may go as far as investigating into the applicant's family background is intended to screen thoroughly any applicant personally and let in only people the company feels can be trusted and who can commit themselves to the company. That brings to the mind the recruitment procedure at Mazda's Flat Rock, Michigan plant. The screening of application consisted of five demanding stages[69] that frustrated many candidates.

In other words, the Japanese company affords time and money for the hiring process. It is therefore not ready to fire people easily and restart the financially costly and time-consuming process again and again. Once hired, the new employee expects the company to keep him for good and he does not feel psychologically prepared to leave the company in order to undergo another trying hiring process in another company. That may shed the light on the fact that in the Japanese company there are almost no mid-career intakes and that the turnover is low, i.e., there are not so many cases of resignation.

1.2.1.2. Continuous training through OJT and job rotation

After the initial training that can be seen as part of a general orientation, an introduction and an initiation to the working conditions, to the work place, to the work practice and routine, new employees of the Japanese company are to remain during their working life under continuous training for different tasks or nature of jobs they have to deal with (Figure 1–1).

The continuous training is carried out through on-the-job training (OJT) which is closely associated with job rotation programs. In the context of lifetime employment, continuous job rotations imply continuous on-site training without which one can not accomplish efficiently his new job. The initial training for new employees is just the beginning of a process that is part of their work-

---

69 J. J. Fucini and S. Fucini, *Working for the Japanese. Inside Mazda's American auto plant.* N.Y.: The Free Press, 1990, pp. 49–65

ing lifelong activities. Job rotation and on-the-job training should be viewed as complementary.

In the framework of the OJT system, one should understand that both the trainees and the trainers are lifetime employees of the same company. OJT conveys thus both the active and passive meaning of giving and receiving the training. As matter of fact, continuous training may take the passing form of being trained when applied to someone who has just been rotated and assigned to a new post and who is being taught how to perform the tasks of his new job. It may also mean training others, teaching the job routines that the trainer masters to the trainees who does not yet.

Besides, there may be classes held for employees either in company training centers or by third parties, such as consulting houses, university professors, schools that give lessons by correspondence.

1.2.2. Seniority based pay/promotion & lifetime employment

As stated earlier, the Japanese company hires people to keep them till they retire and they hire them straight from school. On the other hand, people are willing to stay in only one company until their retirement.

Lifetime employment implies that the worker will grow up chronologically and socially during his working life at the same company. Therefore, it seems quite reasonable that, as the time passes, his salary will rise steadily as he will be occupying higher positions and assuming more responsibility functions. That explains and justifies what is known as the Japanese management's seniority- based system of salary and promotion. In a non-Japanese business environment, as one moves from company to company his earnings and status are commensurate with his present function and/or position that depend in their turn on the length/importance of his past work experience acquired usually at other firms.

Of course, if one has to devote his working life to one and only one company, he should be entitled to an annual incremental increase of salary and should expect to be promoted to higher positions within the company. If such dreams are not fulfilled and it is felt that there are no perspectives for one's dreams to come true, the most likely option left is to move elsewhere. i.e., to quit the company one has been working for.[70]

---

70 A friend of mine who works at K. bank in Wakayama told me that people often resign when they realize that they have no chance to become kacho (section chief)

Figure 1–2 In-house training, mutual commitment and seniority-based system

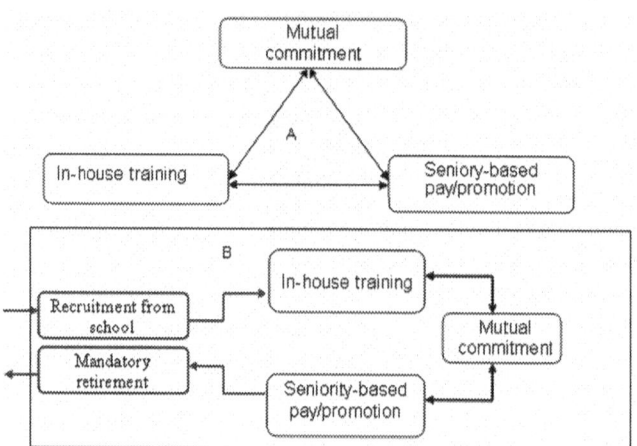

Figure 1–2 A is a traditionally recognized solid triangular representation of the relationship between training, mutual commitment (lifetime employment) and seniority system of pay/promotion. Figure 1–2 B integrates the relationship with recruitment and retirement in the career path of the worker. Furthermore, it shows that training and seniority system are cemented and linked to each other by mutual commitment without which no apparent relationship could be established between training and the seniority system.

Japanese company's management practices are such that persons of the same education who join the company at the same time make up a distinct though informal group within which the cohesion is very strong and group members feel equal. Each group has its own "senpai" and "kohai" (senior and junior), i.e., those who entered the company years before or years later respectively. I am going to refer to such a group based on the year people enter a company as a generation.

The basic pay for employees of the same generation is almost the same even though the nature of work they are assigned to differs greatly and that some of them may be holding higher positions than others. On the average, the basic pay for each generation of workers is lower than that of their "senpai" and higher than that of their "kohai". As regards the relationship between age and education, it should be noted that in Japan, if people have the same level of formal education, there is almost no doubt that older ones must have finished their studies earlier than younger ones.

Ouchi relates an interesting request made to a foreign manager by his Japanese female factory workers:

Can't our plant have the same compensation system as other Japanese com-
panies? When you hire a new girl, her starting wage should be fixed by her
age. An eighteen-year-old should be paid more than a sixteen-year old.[71]

In a non-Japanese context of managements the same idea might have been
expressed as follows: The starting salary should be fixed by the level of education.
A high school graduate should be paid more than a middle school graduate.

Takagi alludes also to the fact that at Toyota people of the same age have the
same pay but of course as he was addressing the Japanese audience, he did not
mention the length of service factor:

(…) as for general pay, there was almost no difference for employees of the
same age.[72]

In order to grasp both the meaning of the young female factory workers'
request and Takagi's thought about that aspect of the Japanese company management a foreign observer should not lose sight of the fact that, in the Japanese
context, formal education is automatically and implicitly implied when speaking
of someone's age.[73]

1.2.3. Lifetime employment and labor management relation

It is difficult, even impossible to imagine a company that features a lifetime
employment whose managers and employees would not be able to trust each others. People can live together for life and share the same values only when they have
smooth relationship. Otherwise, they have to divorce even though they promised
to each others, to their community and/or to their God that they would live
together till their death. Because employees and managers have to share the same
fate that is linked to their company fate, they develop, establish and entertain very
good relations. Following is a testimony by Morita:

The most important mission for a Japanese manager is to develop a healthy
relationship with his employees, to create a family-like feeling within the

---

71 W. G. Ouchi, *Theory Z.* p. 41

72 T. Takagi, *Toyota kara kacho ga kieta. Mono, kane kara hito no kasseika no jidai*
*he.*Tokyo: Gomashobo, 1990, p. 26

73 The question, "How are is your child's old?", one of the most natural answers in
Japans is, for example: "Koko san-nensei" (3rd-year high school student). That means
he/she is eighteen or so

corporation, a feeling that employees and managers share the same fate. Those companies that are most successful in Japan are those that have managed to create a shared sense of a fate among all employees, what Americans call labor and management, and the shareholders.[74]

The company is not only the managers' thing for which workers work, it becomes everyone's thing. This is illustrated in the career path of both the university and non-university graduate.

1.2.3.1. Probable career path of a university graduate

The university graduate, upon entering a Japanese company as ordinary an ordinary worker (hira-shaiin) is going to be in all probabilities a labor union member. But he expects himself to be promoted sooner than later (when compared to a high school graduate), after he has built some seniority, to management posts. Many companies evaluate the workers' ability for promotion through promotion tests and examinations. The school attended and the school records do have some influence on the recruitment. The school records do not exercise much weight concerning considerations for promotion as confirmed in the case of Sony which has established the following rule:

> *Once we hire an employee, his school records are a matter of the past and are no longer used to evaluate his work or decide on his promotion. (...) ... we were seeking ability, not just people with school pride.*[75]

As for promotions, besides the length of service and age criteria, ability is an element that is paid also careful attention to. To well understand this scheme of the career path of a Japanese university graduate, let me contrast it with his foreign counterpart. The Japanese university graduate should assume during his active career many managerial positions and different functions. He is as mobile as his American counterpart but his mobility takes place inside his company. The American university graduate will go probably from company to company while assuming the same kinds of functions or performing similar tasks. In other words, it may seem boring, in other countries, to stay in the same company since one will be working in the same department from the beginning to the end of his career. If one needs fresh air, he has to change companies. In the Japanese company, you change the air by moving from a department to another department, from a section to a different section, etc.

---

74 A. Morita, *Made in Japan*, p. 130

75 Ibid., p. 145- 146

1.2.3.2. Probable career path of a non-university graduate

Most of high school graduates belong to the company labor. The career path of the non-university graduate is somewhat similar though a bit different from that of the university graduate. The former may like the latter reach management posts, even the top management post. I have in mind the case of Mr. Nakao of the Matsushita group and Mr. Ohno of the Toyota Corporation. It should be however recognized that few may climb the management ladder as high as these two. Most non-university graduates spend their entire working life on the production line. Some may emerge as labor union leaders and later be promoted to management posts before retiring.

The possibility for an employee to become a manager one day, and even a member of the board, keeps his dreams alive and gives him the feeling that the company he works in does not only belong to a selected class of people, i.e., managers but that it belongs to him and his peers. This is particularly true when the promotion system does not emphasize too much the formal education background, but takes into account the length of service and the merit of each individual worker. A meritocratic system of evaluation for promotion is felt by every employee as a very fair, particularly when combined with the length of service.

The career path of both university and non-university graduates shows that there is no categorization of employees into two distinct classes, i.e., the ones who can make it to management positions and the others who would remain production operators their career long, as is the case in the U.S. automobile industry. In their career path, a university as well as a non university graduates can reach the top echelons of the company ladder. The main difference being that the non-university graduate may go up his path to the top management by passing through the labor. Another difference resides in the fact that a non-university graduate has to go through much more steps before reaching management ranks.

Labor leaders in most Japanese companies consider themselves potential managers of tomorrow. The management policy of Sony on this point may be taken as a good illustration:

> *In our labor relations, we have a kind of equality that does not exist elsewhere. We see very little distinction at Sony between blue- and white-collar workers. And if a man or woman becomes successful as a union leader, we are very interested, because these are the kind of people we are looking for in our management ranks, people who can be persuasive, can make people want to cooperate with them ... we are constantly looking for capable persons with these qualities, and to rule people out because they lack certain*

school credentials or because of the job they happen to find themselves in is simply shortsighted.[76]

In such environments, the relationship between labor and management can not be but smooth. Besides, the sense of sharing the some fate, i.e., the company fate that ties up management and labor together keeps both parties united and the most likely option left for them is that of cooperation. The company is their company. Their life depends upon the company's prosperity. They feel like the crew members of the same spaceship.

1.2.3.3. Retirement in the lifetime employment system

The retirement offers three paths: a) new employment, thanks to the company recommendation, by subsidiaries or subcontractors; b) re-employment on a part-time basis at the same company; c) just leaving the company with only the financial compensation without any clear work perspective at the horizon.[77]

There is also the retirement imposed by the business law obliging those who get into the board to resign as company employees. This is a theoretical retirement.

1.2.4. Merit, competition and promotion

As one can see it, a university as well as a non-university graduate, a unionized as well as a non-unionized worker can climb up during his lifelong career within one company the hierarchy ladder of that company to become a manager. Either has a chance to become the top executive.

Anyway it sounds in reality as though till the rank of department head, one of the most prevailing criteria is that to make it, besides the seniority, ability, capability and skills play an important role. How to identify those qualities?

At Kubota things go as follows:

There are nine ranks of qualification, and at certain ranks, written examination and interviews are conducted in order to be impartial. A job evaluation and qualification system which is a mixture of seniority and competency is established.[78]

---

76 Ibid., p. 140

77 Contrary to French people who would be happy after retirement and enjoy free time, in Japan being inactive is very hard to stand.

78 T. Kagono, and Kansai Seisansei Honbu (eds.), *How Japanese companies work/Midori ga kaita nihoneki no keiei*, Tokyo: Nihon Keizai Shimbun-sha, 1984, p. 149

At JR, the Japan Railroad Corporation (formerly a state-owned company), the level of education does not make any difference in promotion: only those who pass tests are promoted, be they university or high school graduates and the promotion tests are the same for everyone.

> *But with the privatization, promotion tests based on the ability principle has been introduced. University graduates as well as high school graduates take the same test of ability (there are nine ranks for ordinary employees, five ranks for management posts at the headquarters making in all fourteen test levels). In the past, university graduates were hired by the headquarters for the Shinkansen lines and high school graduates by regional and local directions for ordinary trains: in principle, this difference, according to the (new) regulations, does not exist any more. The decision about who should go where is based on the results of administered tests. Only capable people, university graduates or high school graduates have access to upper posts.*[79]

The higher you go up, the fewer the number of available posts, the fiercer the competition, and the older the people occupying those posts. One should keep present in mind the fact that once someone enters a Japanese company, he is going to stay with the company for his entire working life. His promotion is from within. But the organization chart of the Japanese company is not different from that of the American business enterprise. As one climbs up, the number of posts gets fewer and fewer. At the top, there is just one post for the chairman, one for the president, a few for the board of directors. This implies that the Japanese company should be one of those places where there should be a very high tension, rivalry and competition among managers. Theoretically, the most competent ones go up their way and can even reach the top rank of management hierarchy.[80] Others have to retire at lower ranks. [81]

1.2.5. Lifetime employment, group management and decision making process

As for the relationship between lifetime employment, group management and decisions process, the following points will be considered: (1) workers' involvement in the company management, (2) meaning of the consensus decision, (3)

---

79   Nihon Keizai Shimbu-sha (ed.), *Terase de Nromu nikon no keiei*. 1989, p. 22

80   See the case of Ohno of Toyota who was ejected from the board not because he was less competent but maybe he was politically not supported

81   R T. Pascale & A .G. Athos, *The art of the Japanese management.*, p. 246–247; R. Clark, *The Japanese company*, pp. 115, 116, 202

group management and loose division of work and (4) group responsibility and division of work.

1.2.5.1. Participative management

In the context of group management, the participation of everyone is requested in the form of suggestions for improvement and proposals for decisions.

1) Suggestions for improvement

The group management is based on everyone's participation in and contribution to running better the group, the department and the company one belongs to. This is carried out through suggestions for improvement. At Sony, the reasoning goes as follows:

> *A company will get nowhere if all the thinking is left to management. Everyone in the company must contribute, and for the lower level employees their contribution must be more than just manual labor. We insist that all of our employees contribute their minds. Today we get an average of eight suggestions a year from each of our employees, and most of the suggestions have to do with making their own jobs easier or their work more reliable or a process more efficient.*[82]

Of course, that works only in the framework of lifetime employment where employees and managers feel the need to improve their work place. They cannot look outside for a better work place, say, in another company because there is no hope of finding such a company.

2) Decision proposals

Many decisions in the Japanese system of group management find their origin in proposals from all levels of management. Those at the base have to be trusted because they are going to stay in the company their working life long and are in all probabilities those who will be company top executives tomorrow. Because of that system of lifetime employment, a decision suggested at lower levels cannot be taken lightly. Zimmer goes so far that he thinks the middle management has the monopoly of decisions:

> *The decision making process is the typical business organization tends to be concentrated in the middle management ... (...) decision making is essen-*

---

82 A. Morita, *Made in Japan*, p. 149

tially middle management activity. This differs from the American system, where senior management makes most of the decisions.[83]

Morita explains the importance of getting decision proposals from young managers for the future of the company:

The group management system of Japan, where decisions often are made based on proposal from younger management, can be an advantage for a company. Younger managers can be expected to remain with the same company for twenty or thirty years, and in ten years or so they will move into top management jobs. Because of this the young managers are always looking ahead to what they want the company to be when they take it over.[84]

1.2.5.2. Consensus decision

The decision by consensus is possible and effective only in the context of what it is called group management or participative management. And the latter can work only in the context of lifetime employment because people taking decisions today cannot take them lightly since their future depends on those decisions. That is why the famous consensus about a decision in the Japanese company is time-consuming.[85] But there is a great advantage, i.e., everyone commits himself spontaneously and heartily in implementing a decision agreed upon by the group.

Once a decision is reached, whether it originally came up from the shop floor or down from the from office, it is the Japanese way for everyone to devote every effort to implementing it without the snipping and backbiting and obstruction that is sometimes seen in some Western companies (see p. 40)[86]

1.2.5.3. Group management and loose division of work

In general, the division of work is not so sharp. Group management in its essence means loose work division within the group.

---

83 M. Zimmerman, *How to do business with the Japanese*, Tokyo: Ch. E. Tuttle, 1987, p. 119

84 A. Morita, *Made in Japan*, p. 199

85 Ibid-, p. 198

86 Ibid. p. 199

Any foreigner entering a Japanese restaurant or shop for the first time is stuck by the fact that there is no one stuck to the cash register. Any waiter or shop clerk is ready to serve at the cash register. That case can be contrasted with Flat Rock, Michigan plant of Mazda. American workers could hardly accept and get used to the idea that when a group member falls ill and can not come to work (show up to the work place), his job should be done by other workers, because for them that means not only more work load but principally extra-work for other workers.[87] In the typical American company, a sick person is replaced with someone else of the same skills. Group management presupposes that individual members of the same group trust each others and know each others' tasks or jobs. That is why there is no sharp division of work or task distribution.

1.2.5.4. Group management and collective responsibility

It has been shown that group management implies that the group is held the responsible number one for what has to be done even though individuals have specific tasks to perform. It means that the responsibility for success or failure is a matter of the whole group. In case of success, all the group members indistinctly should be commended. Ouchi reports a case of a Tokyo factory where a foreign boss tried to compensate workers according to their productivity. The group protested against the idea rejecting it strongly:

> *The idea that anyone of us can be more productive than another must be wrong, because none of us in the final assembly line could make a thing unless all of the other people in the plan had done their jobs right first. To single person out as being more productive is wrong and is also personally humiliating to us.*[88]

In the same way, an individual may be at the base of a failure but the group as a whole accepts the blame. Let us listen to Morita:

> *(..) we think it is unwise and unnecessary to define individual responsibility too clearly, because everyone is taught to act like a family member ready to do what is necessary. If something goes wrong, it is considered bad taste for management to inquire who made the mistake ... The important thing in*

---

87 See J. J. Fucini and S. Fucini. *Working for the Japanese*, 1990

88 W.G. Ouchi, *Theory Z*, p. 41

view is not to pin the blame for mistake on somebody, but rather to find out what caused the mistake.[89]

1.2.5.5. Consensus decision and role of the chief

By his personality, the chief instills dynamism, life and strength in the group and all decisions bear his mark. He is the most influential factor for the consensus. Many scholars qualify the Japanese decision process to be bottom-up instead of top-down oriented. The process may be so but the decision is taken at the highest level and the decision once taken by the top executive means the consensus (even though during the debate the consensus has not clearly been achieved as is almost always the case). The following statement by Morita is very explicit concerning this point:

> *[…] and very often the consensus is formed from the top down, not from the bottom-up, as some observers of Japan have written.*[90]

Management by group instead of by individuals implies that the responsibility on the work is collective.

However, if it seems necessary to single out an individual for a failure, it is usually the head of the group, the head of the department or the president who in the name of the group, the department or the company accepts the blame and its consequences. And he does so not as an individual but only as the group representative:

> *Often if some major failure or illegality takes place somewhere within the company, or if there is a breach of trust with the customers, it is the president who resigns to accept responsibility for the failure of the company to do what was correct. Rarely is such an executive personally held responsible for the failure.*[91]

The chief, in a word, identifies himself with the group. That sheds the light on the fact that when in 1985 a Japan Airlines plane crashed costing the life to more than five hundred and twenty people, it was the president of the company who

---

89   A. Morita, *Made in Japan*, pp. 149–150

90   Ibid., 198

91   Ibid., p. 179

resigned from his position to take the blame for the suffering incurred by so many families.[92]

Summary

As it can be realized, Japanese management features should not be considered as unrelated to each others. On the contrary, they depend and imply each others in such a way that an isolated feature of the Japanese management may be quite inefficient. Understanding management traits as scattered elements would consist in missing to grasp Japanese management as a system and that would be very superficial and not useful at all.

The whole structure of Japanese management has been explained and analyzed from the point of view of lifetime employment. Most Japanese management characteristics are logically related to lifetime employment (see Figure 1–3) and would hardly work outside such a system.

---

92   Ibid., 178–179

Chapter 2 Survey-based study of Japanese management in the small and mid-sise manufacturing enterprises

This chapter is going to deal with some management features of the Japanese big company (JBC) described in the first chapter. Its originality consists in the fact that it is based on a survey, though of a limited range, that has tried to investigate the distribution extent of some Japanese big company management (JBCM) features[1] among small and mid-size manufacturing enterprises.

The majority of the researches devoted to the field of the Japanese management focus, from the company scale point of view, only on one kind of corporations, i.e., giant corporations[2]. Some of those well-known "classic" books seem to concentrate on one, two or three big corporations and the drawn conclusions are extended to the Japanese company in general. In that perspective, Pascale & Athos[3] have Matsushita as their Japanese model they keep comparing with the American AT&T; Clark[4] analyzes mainly a company he refers to as Marumaru, which is his Japanese paradigm.

The main point of the present discussion is the question whether it is logical to qualify practices of the dominant minority of a society as the special characteristics of that society. In other words, does it make much sense to qualify features specific to Japanese big corporations as those of the Japanese company in general?

Besides that, most of those scientific researches on the Japanese management are based only on qualitative approaches such as interviews, factory visits, etc. This is an analysis of data collected thanks to a survey.

---

1   According to academic theories, there are four pillars in the concept of Japanese management: lifetime employment, seniority-based system of promotion and payment, cooperative relations between labor and management and consensus decision making process. This chapter limits its range to some aspects of the first three pillars

2   See R.Clark, *The Japanese company*, 1987; W.G. Ouchi, *Theory Z*, 1982; Pascale, R.T. & A.G. Athos, *The art of the Japanese management*, 1982

3   See R.T. Pascale & A.G. Athos, op. cit

4   See R.Clark, *The Japanese company*, 1987

2.1. Survey

In fact, a questionnaire in the form of a check list helped to gather some information on the distribution of what is known as common features of the Japanese management among the manufacturing. By checking a feature, companies confirmed the feature in question to be a part of their management practices. In most cases, a mere "yes" or "no" was the only thing that was needed. Of course, it was not possible to deal with all management features. A choice had to be done, i.e., considering some at the expense of others. Investigated management features are: promotion, salary and allowances, turnover and recruitment, labor/management relation and OJT, job rotation.

2.1.1 Sample size and specs

The survey results are based on data provided by 129 companies that were kind enough to fill out partially or completely and return back the questionnaire to me. (About thirty other companies sent me back a blank questionnaire and they are not included in the 129, of course). In a word, the sample is made up of 129 elements that are all manufacturing companies belonging to the Chamber of Commerce and Industry of Osaka.

Of those companies, only three consider themselves to be large scale manufacturing companies. The others are either medium-size (27.9%) or small (69.8%) firms.

Those 129 companies employ 16171 people. The average is 125 persons by company. One company features the smallest staff of four people and the other extremity is made also of one company employing 1872 workers.

What criteria were used in order to distinguish between small, mid-size and large scale companies? This is another field of research which is completely out of the scope of our concern. The questionnaire asked each company to check its scale: small, medium size or large. 97.7% of the respondents, i.e., 126 companies consider their scale to be small or medium size. We just trusted them and respected their own classification.

More interesting is the question about how we proceeded to select small and mid-size manufacturing companies to be covered by the survey.

The Chamber of Commerce and Industry of Osaka kindly provided us with addresses of its members[5] and no other information than they were all manufacturing companies. As it is known that more than 90 percent of the manufacturing enterprises in Japan are small and mid-size businesses, I did not care about the

---

5   I would like to thank the Chamber of Commerce and industry of Osaka City for their kindness and their help

way to select. I just avoided pharmaceutical and chemical companies when they could be explicitly identified by their names. I just sent questionnaire forms to the first 440 ones listed according to the "a, i, u, e, o" order of the Japanese writing system.

The small number of companies surveyed and the area covered by the survey set limits to the range of the results of the present study. Observations and drawn or arrived-at conclusions concern mainly that small number of companies. One can extrapolate them only with certain reserve to cover all Osaka in the first place. Osaka being the center of the Kansai area, at a second level of extrapolation only, those conclusions can be cautiously applied to all Kansai. Interestingly, the same survey administered (but covering a very few number of companies) ten years later in 2003 in Wakayama Prefecture which belongs like Osaka to Kansai Region confirmed the results obtained in Osaka

If one is sure there exists a Japanese set of management features common to all the Japanese businesses regardless of their geographical location, then and only then should he/she extend those observations to the whole sector of the Japanese manufacturing sector.

2.1.2. How are data presented?

For every item investigated, the number of the sample's elements that dealt with it will be indicated. The group of the respondents will be divided into two sub-groups, that of those confirming to sport the characteristic and that of those who do not. In other words, at the first level, replies will be classified in three categories: Yes, No and NA (not available or no answers). The "NA" category includes both the absence of replies as well as replies that are not valid. Subtracting companies whose reply is classified as NA helps define the range of useful respondents. Where necessary, the percentage of Yes and No answers will be compared. Most of the focus of the present study is on the distribution frequencies of Yes or yes-like replies. Data will be presented mostly in terms of percentages, averages, frequencies of Yes (or yes-like) and No (or no-like) answers within the range defined by the sample of respondents.

2.2. Management features: survey results & management theory[6]

2.2.1. Starting position for management posts

Of the 129 companies, 110 specified the lowest management rank. The starting position for managers differs from company to company. Of the 85.27% of companies that indicated the lowest management position, the rank of group/team leader has the frequency of 44.55% and that of kakaricho, 26.36%. The frequency of the post of kacho, which some academic management theories, hold for the lowest rank for management posts in Japan[7], is considered as such in only 29.09% of the companies. In other words, for the majority of companies (70%), the management starting position is lower than that of kacho. All those findings are shown in Figure 2–1.

**Figure 2–1 Distriburion of companies according to their
lowest management rank**

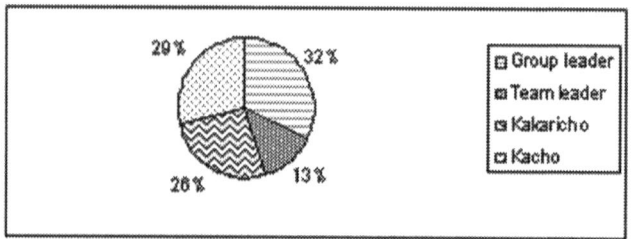

2.2.2. Origin of people holding management posts

96.95% of the companies indicated the provenance of the people occupying management positions. The total number of managers is 2061 for the 125 companies

---

6   For more details about academic management theories on the Japanese management, please refer to J. C. Abegglen, *The Japanese Factory*, 1958 1979 reprint. C. Abegglen & G. Stalk, Jr., *Kaisha, the Japanese corporation*, 1987; T. Uchino, and J.C. Abegglen (eds.), *Tenki ni tatsu nihongata kigyo keiei*, 1988; R. Clark, *The Japanese company*, 1987; K. Odaka, *Japanese management: a forward-looking analysis*. 1986; W.G. Ouchi, *Theory Z*, 1982; K. Urabe,"Innovation and the Japanese management system." In K. Urabe, J. Child and T. Kagono (eds)., *Innovation and management. International comparison*, 1988, pp. 3—25; E. F. Vogel, *Japan as no. 1:. Lessons for America*, 1980

7   See R. Clark, *The Japanese company*, 1987, p.107. However, some scholars consider the rank of kakaricho as the lowest rank for managers. That is what Prof. Uemura's book, *Soshiki no riron to nipponteki keiei* (1982, p.149) suggests though during his seminar he insisted he thinks of team leader or supervisor as the starting positions for management

providing that information. The majority of management posts are filled by people promoted from within (78.5%). And this is true for the top as well as for the middle and lower management. As a matter of fact, among the top management, 81.47% are those promoted from within while for the middle and lower management, 77.6% are so. The number of borrowed and recruited managers represents respectively 7.1% and 6.95% of the posts at the top level of the management hierarchy. As for middle and lower management, there are almost no borrowed managers (1.20%) while people recruited from outside represent a mere 6.43%. The origin of people in management posts is well summarized in Table 2–1.

Table 2–1 Distribution of the number of managers according to their respective origin

Origin	Top managers		Middle/Lower managers		Total	
	Number	%	Number	%	Number	%
Promoted	387	81.47	1231	77.62	1618	78.51
Borrowed	34	7.16	19	1.20	53	2.57
Recruited	33	6.95	102	6.43	135	6.55
Other	21	4.42	234	14.75	255	12.37
Total	475	100.00	1586	100.00	2061	100.00

Is there any explanation beyond these figures? How to understand the fact that the percentage of borrowed and recruited managers is a little higher at the top than at the middle and lower levels of management? As regards a slightly higher average of the number of recruited managers at the top level, two facts may justify that situation. First, big parent or customer corporations small companies deal with usually impose some of their retired managers to sit at the board of directors of their weaker partners, i.e., small and mid-size corporations[8]: they sometimes or very often recommend them for other management posts. Second, the small company may want, of its own will to hire some important retirees of their powerful partners in order to have smooth relationship and easy contacts with their partners' top management[9]. Since a director of a small company can hardly see

---

8   See K. Harada, *Chiisana kaisha no jozuna hito no torikata*, Tokyo: Oesu, 1990

9   See Huddleston.,Jr. J.N. *Gaijin kaisha. Running a foreign business in Japan*, Tokyo: Ch. E. Tuttle Co., 1990

his counterpart of a big corporation[10], hiring a former executive of the parent or customer company will put the small corporation in a better position to deal with the powerful partner[11].

Concerning the borrowed managers, their higher proportion at the top comparatively with the lower level of management may find its explanation in the fact that borrowed managers may have been people dispatched to small and mid-size corporations as the "eyes" of the companies they represent and which are main share-holders[12] or just the financial lending institutions[13]. Such a necessity does not exist at a lower level of management.

That is an attempt of explanation as regards the difference in the number of recruited and borrowed managers among the top and middle/lower management in the manufacturing world of small and mid-size businesses. Anyway, this survey confirms a point of the academic theory that states that the Japanese company promotes people to management posts from within and rarely does it recruit managers from outside. At the same time, it contradicts the management theory stating that promotion from within is a feature specific to JCBM only.

2.2.3. Elements on which depends the starting position for managers

Of the 129 companies constituting the sample of the survey, 119 companies or 92.25% indicated elements that are taken into account for the promotion to the starting post of management. In fact, companies had to check among the following elements all those applying to their respective case: age, education, experience (acquired at another company), special kills/knowledge.

---

10   See R. Clark, *The Japanese company*, 1987

11   The case of Sony Corporation is a very good illustration. At its foundation, Sony included in its board of directors a former top manager of a zaibatsu thanks to whom Sony could sell its shares to secure the necessary working funds(See A. Morita, with E. M. Reingold and NI. Shimomura, *Made in Japan*. Akio Morita and Sony, Glasglow: Fontana Collins, 1987,pp.76–77

12   R. Clark, The Japanese company, 1987

13   J. C. Abegglen & G. Stalk, Jr., *Kaisha, the Japanese corporation*, 1987

Figure 2–2 Distribution of companies according to factors for promotion to management position

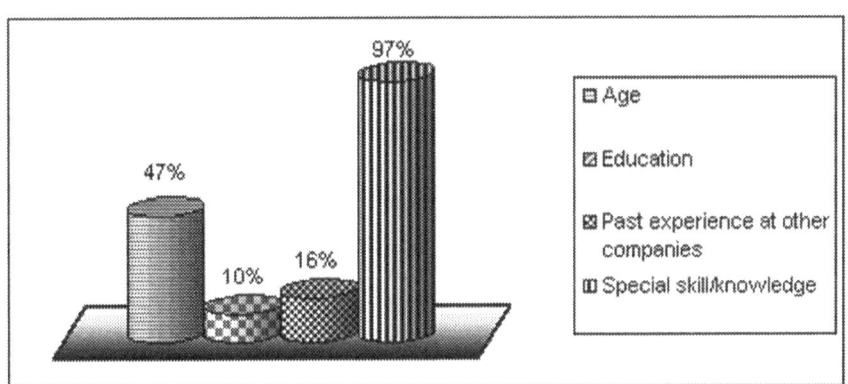

According to the survey results, the most spread criteria among companies for the promotion to the rank of management seems to be special skills and/or special knowledge (all other things being equal). As a matter of fact, special skills and/or knowledge were together rated by 96.6% of the respondents while the rating for the age was 47%. Past experience at/from other corporations ranked third with a frequency of 15.9% and the factor "education" of which the rating stood at 10% is the missing element in most of the companies. The findings about the distribution of elements for promotion are shown in Figure 2–2.

Although the fact that a characteristic is featured by a very large majority does not necessarily imply that it has the highest weight for each company, it remains however very indicative and informative as regards the significance of such of a management trait. Concerning special skills and/or knowledge, the probable importance of its weight relatively to the three others can be seen in the fact that a) the cumulating of their percentage (47.06+10.08+15.97=73.11) is less than the frequency rate of the frequency rate of the factor "skills/knowledge" (96.7%); only 8 companies out 119 or 6.9 % take into account the other three elements together. That implies there are a great number of companies that base their promotion mainly on skills/knowledge.

I specifically asked companies to answer whether a university graduate can start work occupying a management position. Of the 88% of the respondents to that question, a small minority of 10% said "yes". This value is almost identical to the one dealing with education in general (see Figure 2–2).

2.2.4. Promotion factors

Concerning the promotion in general, I proposed some elements to the surveyed companies and asked them to mention all those that are taken into consideration when evaluating a candidate for a promotion. Among factors that weigh when evaluating someone for promotion, the survey tried to check the distribution of age, length of service, ability/competence, special skills/knowledge and education. In fact, I wanted to know the extent to which each of those elements is spread among the manufacturing companies. While academic theories on Japanese management insist on age, length of service and ability/competence as main factors for promotion in Japan, the survey results suggest that in the world of the small and mid-size manufacturing (or at least among the survey sample) ability/competence with a score of 91.7% is the elements to be found in most companies. It may also be one of the most decisive factors for promotion.

It should be noted that the preceding data and those that will be presented below are drawn from the 93.8% of the elements that dealt with the question about the promotion factors.

Figure 2–3 Distribution of promotion factors among companies

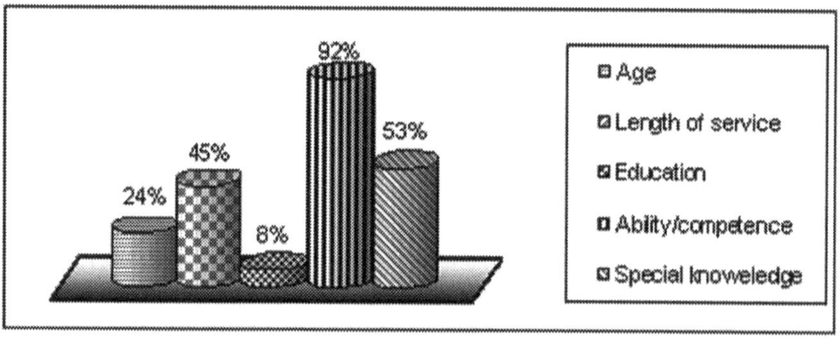

Special skills/knowledge with its score of 52.9% is second while the formal education ranks last with a dismissing 8.3%. Figure 2–3 summarizes the survey results concerning the distributions of factors influencing the promotion.

Why do those companies rate fairly special skilled knowledge and very poorly the education factor? Aren't those two factors closely related to each other? I think that special skills and/or special knowledge in the Japanese business world do mean something earned from experience at the work place and not an academic knowledge like that of someone who has got an American MBA degree.

Therefore, education and special skills/knowledge are not closely related in the Japanese context.

On the other hand, what seems plausible is the combination of ability, competence, special skills & special knowledge in order to make a single set. Using the inclusive logical or of the propositional calculus[14], one may just consider together their combined score of almost one hundred percent (96.7%).

According to the Japanese management theory, age and length of service are closely associated. Here, in the world of the small manufacturing, they seem disassociated as one can see it in the discrepancy of their distribution among companies. Is this due to the fact that employees of the small and mid-size manufacturing are not always hired fresh from school but in the workforce market?

2.2.5. Salary components

The scope of companies that dealt with the question about the elements considered for the salary is made up of 119 companies or 92.24% of the sample elements. That defines also the range of the observations that will follow. It is said that because of the seniority system, the salary of the Japanese worker is generally proportional to his age and the number of years he has worked for the company. In this perspective, the rank of kacho was abolished at Toyota but that should not have affected so much the salary of the former middle management class[15] who lost their titles because those who enter a Japanese company at the same time will almost always have the same salary, though they may have different ranks.[16]

Figure 2–4 Distribution of salary components

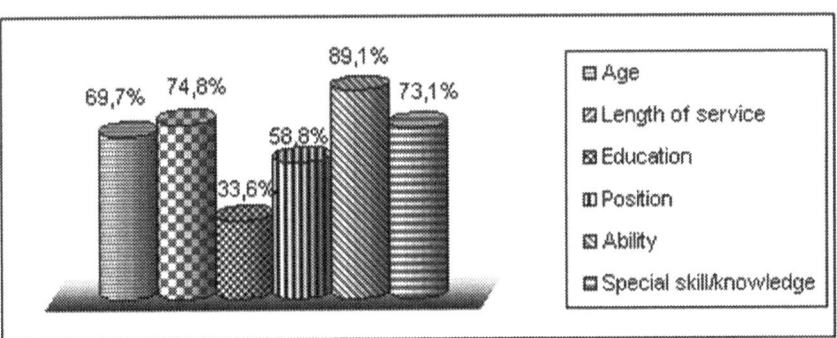

89,1%

69,7% 74,8%   73,1%

58,8%

33,6%

□ Age
▨ Length of service
▧ Education
▥ Position
▨ Ability
▤ Special skill/knowledge

---

14 E. J. Lemmon, *Beginning Logic*, Ontario: Th. Nelson and Sons, Ltd, 1965

15 T. Takagi, *Toyota kara kacho ga kieta. Mono, kane kara hito no ka seika no iidai he.* Tokyo: Gomashobo, 1990

16 See also R. Clark, *The Japanese company*, 1987

The survey results on this point are striking (Figure 2–4). Ability seems to be the most common factor with a rating score of 89.1% on a scale of one hundred. The association of ability with special skills/knowledge (73.1%) has a spreading rate of almost 92.24%. On the other hand, putting together length of service (74.8%) and age (69.74%) covers 86.6% of the companies surveyed.

As one can see it, more than half of those companies surveyed (58.8%) said the position is taken into account, which implies that theoretically the elimination of the middle class of management (as Toyota did) would afflict negatively, the salaries in more than half of the small and mid-size manufacturing world. Therefore, the perspectives of eliminating, in the small and mid- size manufacturing, the middle management class seem unlikely unless salary elements are reconsidered first.

2.2.6. Allowance/reward components

The number of respondents concerning this point is 113 out of 129 companies. That represents 87.6%. The following data are to be situated within the group of those respondents.

According to the survey, the most spread type of allowances is the function or nature of work allowance with a sounding score of 73.5%, followed by special skills/knowledge allowance (48.7%). The length of service ranks third at 41%. Figure 2–5 shows the survey results concerning the distribution of the chosen types of allowances.

Figure 2–5 Distribution of some types of allowances

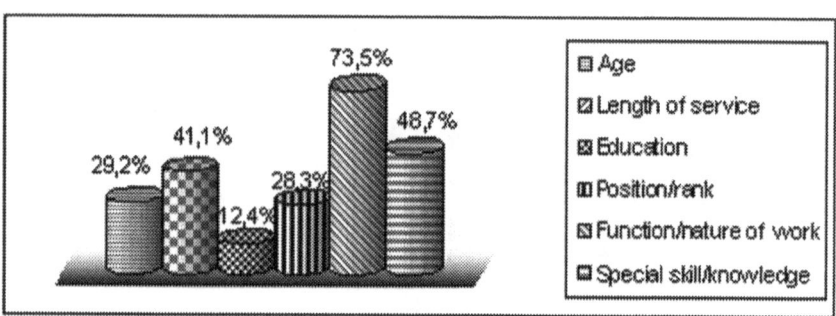

While in the USA, allowances for position and function/nature of work are closely related, in Japan, they are disassociated. Thus, promising young people may be doing the work which should be that of someone of a higher management rank. This way, you may find people having the responsibility of kacho while they are only kakaricho.[17] This aspect of the management theory seems to be confirmed by the survey results that show that most companies pay special allowance for the function but not for the position.

2.2.7. Reasons for leaving the company

Two companies did not deal with the question concerning the turnover. Following are the results about the 98.4% of the remaining companies.

The companies covered by the survey had for the previous year a total turnover of 732 people out of 15952 workers for 127 companies; that represents a turnover of 4.6% and an average of 5.8 persons leaving each company. Does that confirm the management theory according to which the turnover in Japan is very low? I am more interested in the reasons for leaving the company than in the turnover rate itself.

Figure 2–6 Turnover (number of employees frequency in %)

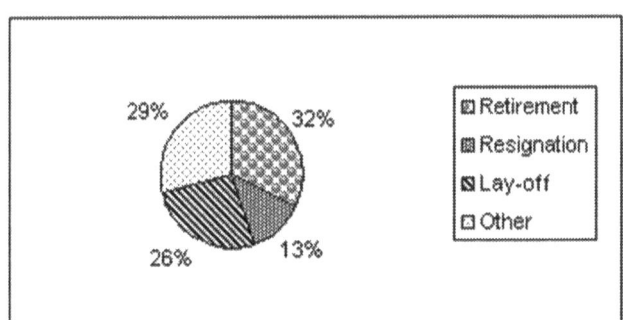

Figure 2–6 shows that more than two thirds of those who left their companies did so because they resigned. This suggests that they did move to other companies, i.e., they changed companies. On the average, two thirds of those who left their companies resigned (66%), 16% did by retirement and 2% were laid-off.

It may be of some interest to point out that in the small and mid-size manufacturing, attitudes of the company and those of the workers are quite different. The small and mid-size company tends to keep its employees as long as possible while the latter ones seem not to commit themselves to the company. This difference of attitudes may shed some light on the very high rate of resignation on the one hand, and on the other on the very low rate of lay-off.

---

17   Ibid.

2.2.8. Recruitment

The number of the respondents represents 98.4%.

The Japanese company is known for recruiting young people fresh from school, who will work for that company all their work life (-some scholars suggest that the commitment of the Japanese worker to his company is what makes the Japanese worker's productivity so high).

Figure 2–7 Distribution of companies according to recruitment modes

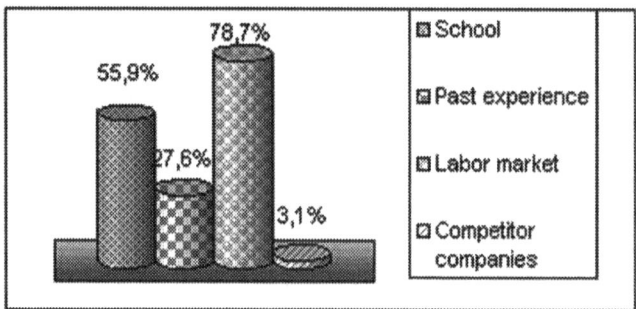

However, the survey (Figure2–7) reveals that small and mid-size manufacturing enterprises recruit mainly in the work force market.[18] This element is mentioned by 78.7% of the respondents and recruitment from school ranks second at 55.9%. Few companies mentioned attracting people from competitor companies. Only 27.6% of the companies surveyed mentioned past experience at other companies as an element they also take into account while recruiting. Recruiting competitors' employees seems to be out of the target range of the management strategies of most companies (3.1% of companies only are interested in doing so).

2.2.9. Labor/management

The following findings reflect the labor/management relationship at 116 companies out of 129, i.e., 89.9% of the sample elements.

The collected information on the labor/management relationship reflects perfectly the management theory according to which the relationship is that of cooperation, collaboration and mutual understanding.[19] These three elements

---

18   Only one company says it would like to recruit from school but graduates are attracted by and go to big corporations

19   K. Urabe "Innovation and the Japanese management system." In K. Urabe, J. Child and T. Kagono (eds.), *Innovation and management. International comparison.*, 1988, pp. *3–25*

when understood from the view point of the math set theory or the propositional calculus method, their union will yield a rating score of 99.13%. That is to say that 115 out of 116 mentioned at least one of the three types of relationship. The bargaining relationship between management and labor is present in only 18.1% of the respondents. Figure 2–8 shows in details the survey results.

There are some unionized workers but their number is insignificant and about all of them are members of their respective in-company labor union. 44.2% of companies dealt with the question about the existence of labor unions, and of that minority, 50% mentioned to have their own labor unions.

Closely relating to this feature is the quasi-absence of strikes (of the 95.3% of respondents, 6.5% experienced strikes) and that of strike threatening (of 42% of respondents, there were strike threats at 16% of the companies) for the last ten years.[20]

Figure 2–8 Distribution of companies according to the nature of labor/management relationship type

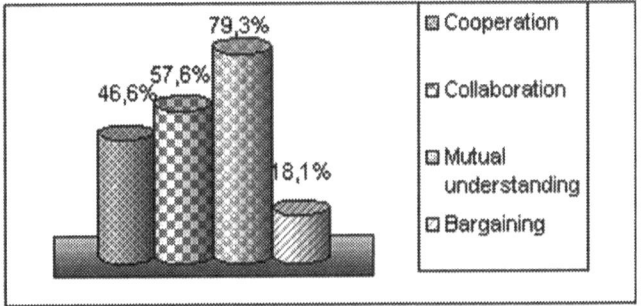

A very small minority of companies representing 28.7% of the sample informed whether or not a member of the labor union can become a manager. Within that very small minority, a very large majority of 83.8% confirmed the fact that ancient labor union members have become or may become managers.

One thing has struck my mind. Even those companies that said to have not any kind of unions at all confirmed however that the relationship between management and labor union is based on the spirit of collaboration, cooperation and mutual understanding. I think this has to be understood as follows. In the Japanese context, the company labor union is so cooperative (a foreign observer would say: very weak) that there is almost no difference between officially unionized workers and non-unionized ones. Therefore, one can now understand why those companies, though without any labor union at all, mentioned to have excellent relationship with such organizations. Labor unions in the latter case should

---

20   One company has experienced once a strike and a strike threatening

be interpreted as the total work force no matter whether they belong to a declared labor organization or not.

2.2.10. Within-the-company competition

In big corporations, small groups and individuals emulate[21] through the well-known QCC and suggestion systems.[22] However, in the world of the small and mid-size manufacturing, the competition within the company is mentioned as encouraged by less than half (40.16%) of the very large majority of the respondents (94.6%).

Figure 2–9 Distribution of companies encouraging internal competition

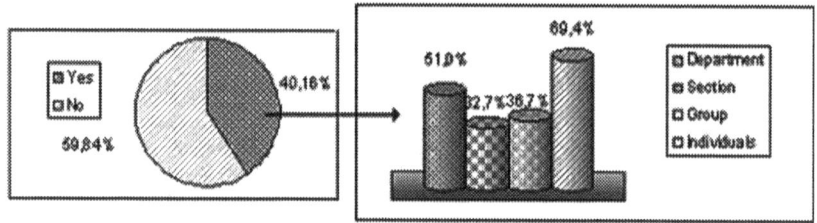

Among the 59.6% of the companies that specified the place where they wish the competition to take place, 32.5% mentioned departments, 20.8% indicated sections while 23.4% checked groups preferred competition instances. As regards the competition between individuals, of the 60.5% that dealt with the question, a large minority of 43.6% encourage the competition at the individual level (Figure2–9).

From a different but complementary observation point, it can be noted that among the 49 companies out of 122 respondents (40.16% of 94.6%), the competition is promoted at department level in 25 companies (51%), at the section level in 16 companies (32.7%), at the group level in 18 companies (36.7%) and in 34 companies (69.4%) at the level of individuals. It is interesting to notice that the two highest frequencies represent the bottom (individual level) and the top (department) of the company organization involved in the production process (see Figure 2–9).

---

21 Sh.Uemura, Soshiki no riron to nipponteki keiei, Tokyo: Bunshindo, 1982, pp. 172–173

22 Y. Monden, R. Shibakawa, S. Takavanagi and T. Nagao (eds.), *Innovation in management. The Japanese corporation*, Norcross, Georgia: Industrial Engineering and Management Press, 1985; Sh. Shingo, *Non-Stock Production. The Shingo system for continuous improvement*, Cambridge, Mass.: Productivity Press, 1988

Things should be made clear here. I did not want to know if the competition exists within the company and at which level. But I wanted to know whether the competition is encouraged and may then be considered a strategic force or element of the company policy

2.2.11. OJT and Job Rotation

Job rotation (JR) and on-the-job training (OJT) are two complementary practices of Japanese management. The number of respondents concerning the job rotation represents a very large majority of the sample elements with a frequency of 97.7%. Among that large majority, only a minority of 41.3% mentioned to have the system of rotating workers. In other words, the job rotation is practiced by 52 companies out of a total of 126. Those findings are shown in Figure 2–10.

Figure 2–10 Number of companies practicing JR

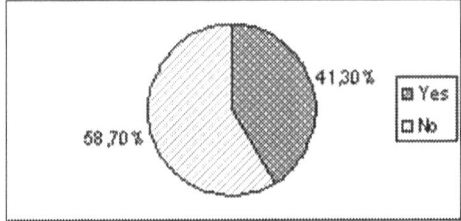

Concerning OJT, the range of respondents represents 96.9% of the sample elements. Of those respondents, a strong majority of 64% have on the-job-training practices for their workers as one can see it in Figure 2–11

Figure 2–11 Number of companies with OJT program

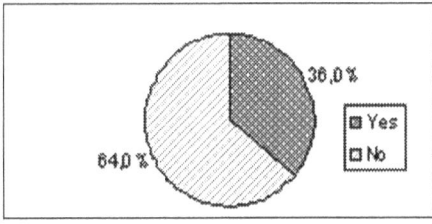

The survey tried also to know the kind or the status of workers concerned with the OJT programs. 30% of the 80 companies conduct such programs for managers, 41% for office workers and 11% for line operators. For 55% of the companies, OJT involves everybody. It is worth noting that those categories of replies are not exclusive of each others. i.e., some companies mentioned more than one category of workers who are targeted by OJT.

In 42% of the cases, OJ'T is voluntary, in 57% it is linked to the nature of work or the function of the worker and in 28.7% of the companies, OJT programs are conducted/implemented for the purpose of promotion.

2.2.12. Bonus system

The survey confirmed the fact that the payment of the bonus system is a characteristic of the Japanese company in general.

All the surveyed companies pay a bonus to their employees.

The mode is five months with a frequency of 42 (companies) for the minimum and 40 (companies) for the maximum. The average varies between 5.27 and 5.36 months. In most companies, the amount of the bonus ranges between 2 months and 6 months (see Table 2–2).

Concerning the frequency of the bonus payment twice is the standard with a frequency of 122. Six companies pay the bonus three times a year Data are not available for one company only.

After getting knowledge of the survey results, what to think of the Japanese management?

Table 2–2 Distribution of companies according to the volume of bonus (in months of salary)

Bonus value (months of salary)	Minimum value of bonus			Maximum value of bonus		
	companies	Frequency	Cumulative frequency	Number of companies	Frequency	Cumulative frequency
NA	3	2,3%	2,3%	3	2,3%	2,3%
1	2	1,6%	3,9%	1	0,8%	3,1%
2	15	11,6%	15,5%	10	7,8%	10,9%
3	23	17,8%	33,3%	17	13,2%	24,0%
4	25	19,4%	52,7%	22	17,1%	41,1%
5	42	32,6%	85,3%	40	31,0%	72,1%
6	15	11,6%	96,9%	26	20,2%	92,2%
7	1	0,8%	97,7%	7	5,4%	97,7%
8	1	0,8%	98,4%	1	0,8%	98,4%
10	1	0,8%	99,2%	1	0,8%	99,2%
12	1	0,8%	100,0%	1	0,8%	100,0%
	129	100,0%		129	100,0%	

2.3. Necessity to re-think the concept of Japanese management

There is a style of management specific to Japanese company. This is a fact nobody can deny. However, I think that the concept of Japanese management needs some clarification.

From the view point of logic as science of reasoning,[23] a concept may be studied according to its extension and/or to its intension. The extension of a concept is made up of all the individuals or objects to which the concept may apply whereas its comprehension or intension refers to the characteristics of that concept.

As for the Japanese management concept, its extension is supposed to refer to all Japanese companies: small, mid-size and large scale corporations. But the survey reveals that many of the features known as specific to the Japanese company do not apply to all the types of Japanese corporations. As a matter of fact, small and mid-size manufacturing enterprises seem to possess some characteristics that are their own and that are sometimes literally opposed to what is referred is referred to as Japanese management traits, i.e., management of big corporations

Though the style of management specific to the Japanese corporation is a fact no one could doubt of, the content of that concept seems not to reflect the reality. How can one then accept the so-called management features of the Japanese style whereas most companies do not have them? Hence the concept needs re-thinking.

Do the Japanese small and mid-size businesses play an important role in the Japanese manufacturing industry? Are they worth being paid any special attention to?

If the concept of population ecology by organization theory[24] can be of some usefulness, Japanese small, mid-size and large scale companies make up the population of the Japanese manufacturing world.[25] From that point of view, the small and mid-size manufacturing businesses represent more than 90% of the manufacturing companies. That fact is confirmed by the survey: in fact, 126 out of 129 companies or 98.7% selected randomly are part of the small and mid-size enterprises). Second, according to official statistics, 99.5% of the manufacturing world population are small and mid-size manufacturing companies.[26] Furthermore, according to Sakai, Howard and Smitka, the small/mid-size manufacturing is the backbone of the Japanese success and the source of its competitive power in the international market.[27]

---

23   This goes back to the works of Aristotle

24   A. Grandori, *Perspectives on organization theory*, Ballinger Publishing Company, 1987

25   R. Clark (*The Japanese company*, 1987) prefers the phrase "society of industry"

26   Chusho Kigyo Cho (ed.), *Zu de miru chusho kigyo hakusho* Tokyo, Okura-sho, 1092

27   R. Howard, "Can small business help countries compete", Harvard Business Review, Nov/Dec. 1990, pp. SS-103; K. Sakai, "The feudal world of Japanese manufacturing",

On the other hand, when one considers that matter from the workforce point of view, the small and mid-size manufacturing employs three out of four people, i.e., the majority of the working or active population is made up of individuals working in the small and mid-size manufacturing sector.[28] In the USA, the small and mid-size manufacturing businesses represent, according to Howard, only 35% of the employment.[29] That is to say that from this point of view, the Japanese the small and mid-size manufacturing world can not be neglected. Besides that, the human resources are considered one of the most important and critical elements that lie behind the success of the Japanese company.

Furthermore, from the financial point of view, the small and mid-size manufacturing can not be neglected at all. According to the Agency for the Small and Mid-size Enterprises of the Ministry of Finance, the value added to the manufactured products and the shipment value (both values evaluated in yen) by the small and mid-size manufacturing enterprises with a work force inferior to 300 people represents respectively 55.5% and 51.8%. And if companies with the work force not exceeding 1000 people are to be included, the figures rise to 75.5% and 73.4% respectively.[30]

Taking into account that multiple aspects (number of companies, quantity of jobs offered, value-added, volume of shipment) of the small and mid-size enterprise importance within the manufacturing sector, one would be inclined to think that features specific to that group should be considered the dominant traits of the concept of Japanese management. Unfortunately, the reality is quite different.

Re-thinking the concept of Japanese management would consist in distinguishing (Table 2–3): (1) features specific to the majority of big corporations; (2) features specific to the majority of small and mid-size manufacturing; (3) features common to both (1) & (2).

Harvard Business Review. Nov./Dec. 1990, pp. 38–51; M. J. Smitka, Competitive ties. Subcontracting in the Japanese automotive industry, N.Y.: Columbia University Press, 1991

28   Chusho Kigyo Cho (ed.), 1992

29   Howard, R., "Can small business help countries compete?" *Harvard Business Review*, Nov./Dec. 1990, pp. 88–103

30   ChushoKigyo Cho (ed.), *Zu de miru chusho kigyo hakusho*, 1992, table appendix, pp. 8–9

Table 2–3 Distribution of management features in the manufacturing sector

Management Characteristics	(1) Manag. theory	(2) Survey Results		(3) Common Characteristic?
		Frequ. (in %)	Tentative conclusion	
Management starting position				
• Kacho	OK[31]	29.1	--	--
• Lower than kacho	?	70.9	OK	?
Promotion to management factors				
• Age	OK	47.1	--	--
• Education	--	10.1	--	OK
• Experience	--	17.0	--	OK
• Special skills/knowledge	--	96.6	OK	--
Origin of Management people				
• Promoted from within	OK	78.5	OK	OK
• recruited	--	6.6	--	OK
Promotion factors in general				
• Age	OK	27.3	--	--
• Length of service	OK	44.6	OK	OK
• Education	?	8.3	--	?
• Ability/Competence	OK	91.7	OK	ok
• Special skills/knowledge	--	52.9	OK	--
Salary components				
• Age	OK	69.7	OK	OK
• Length of service	OK	74.8	OK	OK
• Education	OK	33.6	--	--
• Ability/competence	?	89.1	OK	?
• Special skills	--	73.1	OK	--
• Position	OK	58.8	OK	OK

---

31 Some specialists of the Japanese management think that management position starts at a lower rank than that of kacho (see footnote 6)

Table 2–3 Distribution of management features in the manufacturing sector (continued)

Management Characteristics	(1) Manag. theory	(2) Survey Results		(3) Common Characteristic?
		Frequ. (in %)	Tentative conclusion	
Allowance/Rewards elements				
• Age	OK	29.2	--	--
• Education	OK	41.1	--	--
• Length of service	OK	12.4	--	--
• Function/nature of work	OK	73.5	OK	OK
• Special skills/knowledge	--	48.7	OK	--
• Position	OK[32]	28.3	--	--
Reason for leaving the company				
• Retirement	OK	16.5	--	--
• Resignation	--	66.1	OK	--
• Lay-off	--	2.05	--	OK
Recruitment				
• From school	OK	55.9	OK	OK
• In the workforce market	--	78.7	OK	--
• Past experience considered	--	27.6	--	OK
• From competitors	--	3.1	--	OK
• Labor/management	•	•	•	•
• Collaboration/cooperation	• OK	• 99.1	• OK	• OK
• Bargaining/adversity	• --	• 18.1	• --	• OK
• In-company-competition encouraged? Yes	OK	40.2	--	--
• Bonus for-every-one system? Yes	OK	100.0	OK	OK
• OJT	• OK	• 64.0	• OK	• OK
• Job rotation system	• OK	• 41.3	• --	• --

The column (1) "Management theory", represents the extension of the Japanese management concept according to current academic theories that usually concentrate on the big corporations. In other words, the Japanese manufacturing world is represented by only 0.5% of its population.

The column (2), "Survey results", is the summary of the field research. It is worth mentioning that only features with a rating score equal or superior to 50%

32   This is according to the late Professor Uemura, an authority in the Japanese management (it is called shoku yaku teate).

are suggested as characteristics of the small and mid-size manufacturing. This is really the forgotten area in the modern management theories on the Japanese company.

The third column (on the right) shows elements common to both big and small/mid-size corporations. And I think that this column contains features that cover completely the extension of the concept of Japanese management. If one decides to consider features contained in the column of "Management theory" as those of the management of the Japanese style, he/she should state clearly and convincingly the reasons. As matter of fact, the concept of Japanese management as defined by academic theories seems not to apply to most corporations. In its present formulation, the extension of that concept is really small but it is presented as though it covers most business enterprises. After this survey, it would be difficult for me (as for any critical mind) to accept the concept of Japanese management as formulated by modern management theories.

2.4. Conclusion

This chapter objective was not to try to discover new features of the Japanese management. Rather, it attempted to check how wide the so-called special features of the Japanese management are spread among Japanese companies. In other words, it endeavored only to check the extension of the concept of Japanese management among the small and mid-size manufacturing companies.

The main discovery here remains the fact that the concept of Japanese management, if it is to be understood as the one that covers the whole population of the Japanese manufacturing sector, appears to have fewer characteristics than academic theories pretend. It obeys the logical rule according to which the wider the extension of a concept, the smaller its comprehension, i.e., distinctive features. Till now the concept of Japanese management has been defined as very rich. That is true because its extension was limited to some big corporations. By expanding its extension to include small and mid- size businesses, I have realized that its features should be reduced.

Why is the concept of Japanese management used in its narrow sense to cover all the population of the manufacturing industry? As everyone has realized, it is not because of the number of companies nor because of the number of jobs offered by big corporations. Is it because of the financial outputs? I am not sure. I rather think it is because big corporations in Japan mean: manufacturers, traders and exporters/importers.

Therefore there is no doubt that the foreigner knows only big corporations. As the concept of Japanese management was first thrown into the academic world by a foreigner,[33] there is no surprise if the concept is what it is.[34] Of course, foreigners are attracted by big corporations that fascinate them abroad and they come to Japan to study those corporations that represent Japan in the international market as well as in the national market.

On the other hand, big corporations in Japan, though numerically not so an important group, are however the policy makers of the Japanese manufacturing industry and the representatives of the Japanese manufacturing world in the international market. They are for the society of the manufacturing industry what the Japanese government cabinet members (ministers), Diet members and other policy makers are for the Japanese society.

However, studying the Japanese society does not mean the study of the governing body only. The same reasoning holds for the manufacturing world. That implies that traits specific to the small group of companies should be included in the characteristics set of the Japanese company in general.

By the way, it would be an error for a foreigner to concentrate all his/her energy on big corporations only. Big Japanese corporations are not exactly what they appear to be. Writes Mr. Sakai,

> *In reality, these huge businesses are more like trading companies. That is rather than design and manufacture their own goods, they actually coordinate a complex design and manufacturing process that involves thousands of smaller companies, The goods you buy with a famous maker's name inscribed on the case are seldom the product of that company's factory- and often not even the product of its own research. Someone else designed it, someone else put it together, someone else stuck it in a box with the famous maker's name on it and then shipped it to its distributors.[35]*

What Sakai means is that the small and mid-size manufacturing is the heart of the Japanese manufacturing.

I hope Japanese scholars will reconsider more carefully the concept of Japanese management in order to provide anyone interested in the Japanese company with the correct notion and accurate information of the Japanese management.

---

33   See J. C. Abegglen, *The Japanese Factory*, 1979

34   Odaka (1984) refers to that as one of what he calls "foreign-born myths' about the Japanese management (K. Odaka, *Japanese management: a forward-looking analysis*, Tokyo: Asian Productivity Organization, 1986 pp. 1–5)

35   K. Sakai, "The feudal world of Japanese manufacturing", Harvard Business Review, Nov./Dec. 1990, pp. 38–51, p.39

Theories on the Japanese management should distinguish that concept in its general and narrow/particular meanings. The particular should not be taken for the general (meaning).

Anyway, being consistent with myself, I have dealt with the survey data as if the three big corporations did count. What is the influence of an element of which the frequency represents only 2.3%?

What did I learn from the analysis of the survey data? I think many assertions about the Japanese management are too sharp and need to be softened. For example statements such as: 1) Japanese management is the management of big corporations and 2) there is a clear cut line between the management style of the big corporations and that of small companies should re-examined.

As we see, the survey results recognize some differences but bring also to the light some areas common to all Japanese companies regardless of their size or scale. A new approach to the concept of Japanese management should distinguish management features that are specific to only (a) big corporations or small/mid-size companies from those that are common or can apply to both (a) and (b).

I nevertheless remain conscious of the limits of this study taking into consideration the small number of companies covered by the survey. If the results of this survey may be trusted, it is because of a striking uniformity of the behavior of the Japanese society in general. From Kyushu to Hokkaido, one gets the impression of hearing the "same" announcement voice on the bus. In the same way, one should expect to see similar ways of managing Japanese companies of the same size and same industry from north to south and from east to west.

Part II JIT/Lean Production Methods

Chapter 3
JIT/Lean production system

Chapter 4
Survey-based study of JIT techniques in the small and mid-size
manufacturing enterprises

Chapter 3 Just in time (JIT)/Lean Production System

3.1. JIT and lean thinking

An analysis of the Just-In-Time (JIT) system reveals that it is a set of production techniques and methods aiming at a) eliminating completely all possible wastes and their causes and, b) reducing drastically the production lead time so that demand and delivery of products can be met in time, making useless the necessity of keeping stocks of finished and/or work-in-process products. JIT aims thus at the non-stock production,[1] and by the same token at the lean production. In fact, it is a lean production method. Furthermore, lean production and JIT are often, and for many, considered synonymous. In a sense, that is true. But there exits however some substantial difference between JIT and lean approaches to production.

JIT emphasizes the necessity of identifying and then eliminating (or drastically reducing) all kinds of wastes relating to the production activities. The outcome is a waste-free production, i.e., a lean production. Of course, for the time being the lean concept and that of JIT seem interchangeable. But if one looks closely and carefully into the matter, he would in fact realize that the concept of lean is broader. We might well some day come out with some different methods of eliminating wastes that could as well result in the lean production. Until then

---

1   Shingo's writings clearly show that TPS is a production system and it aims at zero stock, but the English translation is not clear at all. In fact, the title of his first book about TPS is "Toyota Seisan Hoshiki no IE-teki kosatsu: non stock seisan he no tenkai" that literally translates into English as "Study of Toyota Production System From IE viewpoint: Development Toward a Non-Stock Production". Unfortunately, for reasons that no one knows, the published English translations by Japan Management Association (1981) and Productivity Press (1989) have omitted the subtitle. The subtitle is very important since it became the title of a more important book: "Non-Stock Seisan Hoshiki He No Tenkai: Toyota Seisan Sisutemu No Shin No Igi" that should translate as "Development toward Non-Stock Production: The Real Meaning of Toyota Production System". Unfortunately, the official translation reads: "Non-Stock Production" and subtitle chosen by Productivity Press (Publisher) is "The Shingo System for Continuous Improvement".

and for the time being, only JIT has been clearly identified and recognized as a lean method. That is why it is used as synonymous of the lean production system. Besides, the fact that the lean concept has been illustrated mostly by examples of successful JIT implementation cases has contributed to that impression.

We think however that the lean concept keeps its doors open to any other kind of production system to be qualified as such provided that it results, like JIT, in the lean production. In spite of JIT being for the moment the only clearly defined lean production method, it should, however, be considered *a* lean production method; but surely not *the* lean production method. In other words, the JIT system is *a lean subsystem*, but *not the lean system*. Mathematically speaking, JIT is included in Lean. Unless some day some evidence shows that Lean is a set that can be made up of only one sub-set, i.e., JIT, the two systems should be considered, until then, to feature the relationship of inclusion instead of equality (See Figure 3–1).

Besides, the concept of Lean can not be reduced to a production system only, even though it is JIT production system. Lean does not have and has not yet been able to define its own specific and intrinsic techniques that, once applied, would lead a production to becoming lean. The lean concept is broad and can be applied to a production method, to a production unit, even to an entire enterprise of any industry.[2]

Unlike JIT, Lean is more than a production method; it is rather a characteristic that can be applied to qualify a production method as lean (JIT) or not lean (traditional American model of mass production). Whereas JIT from its inception, by its nature and essence, is and remains a production method, the lean concept was thrown in the academic and business worlds as the result or conclusion of a survey.

> *[The Machine That Changed the World] ... " presented a wealth of benchmarking data to show that there is a better way to organize and manage customer relations, the supply chain; product development, and production operations, an approach pioneered by the Toyota company after World War II. We labeled this new way lean production because it does more and more with less and less.*[3]

Although the fathers of the lean concept, Womack and Jones, seem to equate Lean and JIT, we have sound reasons to stick to the idea that the range of Lean

---

2   We can refer to a manufacturing company as JIT company, but the concept would hardly apply to service industry enterprise.

3   J. P. Womack and D. T. Jones, *Lean thinking*, NY: Simon & Schuster, 1996, p. 10

should be understood as being broader than that of JIT which is just a method among many other possible lean methods, known or not yet identified, virtual or actual, present or future, that may lead to the lean production (See Figure 3–1). But for the time being, dealing with Lean refers undoubtedly to JIT methods.

Whereas JIT emphasizes the thorough elimination of muda, the lean thinking *"provides a way to specify value, lineup value-creating actions … "*[4] Womack and Jones identify five principles of the lean thinking: value, product value stream, value flow, pull of value by customers, and perfection pursuit[5] Value is defined *"in terms of specific products with specific capabilities offered at specific prices through a dialogue with specific customers."*[6]

Figure 3–1 Universe of lean methods

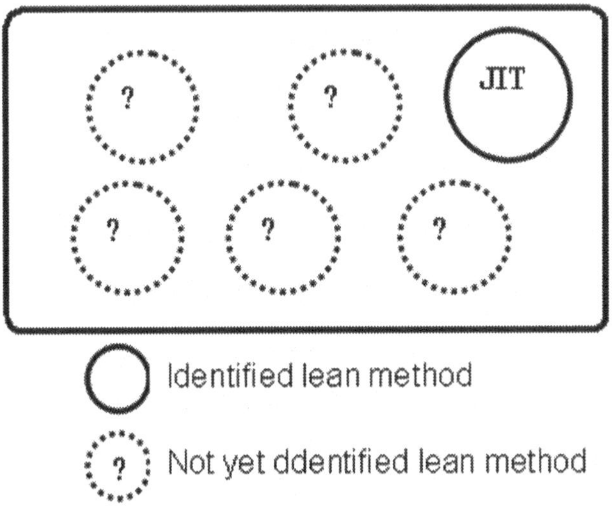

The lean thinking is presented as the antidote to muda[7], the target of the JIT.

JIT and Waste identification

According to Ohno,[8] the movement of the workers can be divided into pure wastes and production operations. And production operations themselves are made

---

4   Ibidem

5   Ibid. p 10

6   Ibid., p 19

7   Ibid. p. 15

8   Ohno T., *Toyota production system*, Productivity Press, 1988, p. 57–58; Ohono, T; *Toyota seisan hoshiki*, 1978, pp. 102–103; Japan Management Association, ed. *Toyota*

up of value-added and non-value-added operations (see Figure 3–2). Activities such as waiting or idle time for workers, stacking work-in-process inventories, double transportation, etc. are pure wastes. Non value-added operations can be considered as wastes, but they are thought of as necessary in order to perform value-added work. Walking to pick up parts, opening packages of parts from suppliers, pushing buttons are examples of non-value-added operations. Value-added operations are processing activities such as assembling, changing the shape or altering the characteristics of a product.

Figure 3–2 Analysis of operator's work-related movement

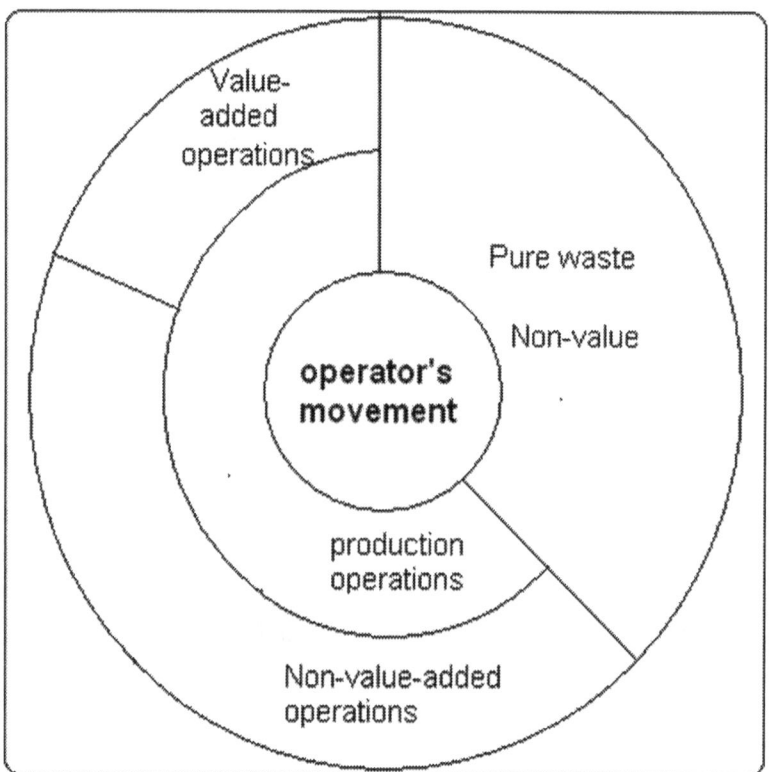

Source: Ohno, 1988, p.102; Japan Management Association, 1978, p. 207

As regards the production, the worker's production operations can tentatively be classified or qualified as relevant (useful) or not relevant; necessary or not necessary. Value added activities are both necessary and relevant to the production; pure muda or wastes are neither relevant nor necessary (see Figure 3–3)

genba kanri, 1978 pp. 206–207

Figure 3–3 Characteristics of worker's movement

	Necessary		Not necessary
Useful or relevant	*Net Value-added or lean operations:* • Changing shape or character of a product • Assembling • Forging • Welding • painting		XXXXX
Not useful or not relevant	*Non-value added operations or Wasteful operations:* • Walking over to another location to receive parts • Removing wrappers from parts purchased from subcontractors • Removing parts in small quantity from a large pallet • Handling a push button already in position		*Pure wastes:* • Time on hand • Meaningless transport • Stockpiling of intermediate products • Changing hands • Transporting to a place other than the destination

As we have just learnt from Ohno, any activity or factor that does not add value to the object of the production (i.e., the product being made or the service being created) is a waste.[9] In the manufacturing, wastes are mostly related either to the worker's or process' operations.

In order to identify different wastes relating to production activities, the following lead question should be applied to every step of the production process or operation: "Does the production activity (or production-related activity) add value to the object being produced?" If the answer is "No", the activity, regardless of its importance or criticality, should be classified as a pure waste to be eliminated since it only raises the production cost.[10]

JIT, as initiated by Ohno recognizes seven categories of wastes:[11]

(1) Waste of overproduction;

(2) Waste of time on hand (waiting time);

(3) Waste transportation;

---

9   See Ohno, T., *Toyota Production System. Beyond large-scale production*, Cambridge, Mass.: Productivity Press, 1988, p.54; Engwall, R. L., "The expanding role *of* IEs in manufacturing', *Industrial Engineering*, June 1989, pp. 52–53

10 See Sh. Shingo, *Study of Toyota Production System from industrial engineering point of view*, Tokyo: Japan Management Association, 1981

11 See T. Ohno, *Toyota Production System. Beyond large-scale production*, 1988, pp. 10–20, p.120

(4) Waste of processing (over-processing or under processing
(5) Waste of stock on hand (inventory);
(6) Waste of movement;
(7) Waste of making defective products.

Analysis of the relationship between the different kinds of primary wastes

Since the different wastes relating the production activities have been defined by Ohno, no one has tried to show the extent to which they are related to each others, i.e. they imply others, or each others. Table 3–1 and Figure 3–4 show that overproduction has some kind of direct relation with the majority of the mudas. It outstands as a clear cause of excessive inventory and operator's motions; indirectly as the cause of transportation.

If you wait, you will not overproduce. If you overproduce, you will not wait. In other words, either you over produce or you wait. Unfortunately, most of the time, people do not wait. They prefer continue producing, i.e. overproducing.

Table 3–1 Matrix table showing the relationship between the different types of muda

	OVE	INV	TOH	OUP	DEF	TRA	MOT	Directly Related to % of the other wastes
OVE		→	←→	←	←		→	83.3%
INV	←					→	→	60.0%
TOH	←→							16.7%
OUP	→				→		→	60.0%
DEF	→			←				33.3%
TRA		←					→	33.3%
MOT	←	←		←		←		66.7%
	83.3%	50.0%	16.7%	60.0%	33.3%	33.3%	66.7%	

Key: →: Cause-effect relationship; ←: Effect-cause relationship; ←→: This applies only to time on hand/overproduction relationship: overproduction hides the waste of time on the hand; and the hiding of the time on hand leads to overproduction

Ove = overproduction; INV = inventory; TOH = time on hand; OUP: over-/underprocessing; DEF= defective; TRA = transportation; MOT = motion

Figure 3–4 Cause and effect relationship between the different types of mudas

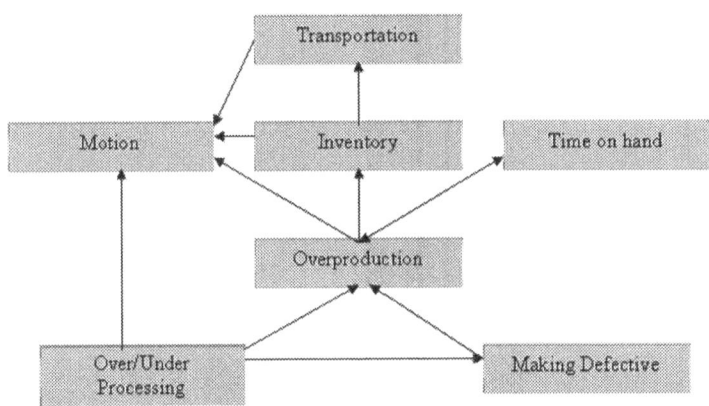

Overproduction, not inventory, is the worst enemy

Interestingly, it is not inventory that Ohno presents as the worst enemy. According to Ohno, inventory is the greatest of mudas but overproduction is the worst of the all wastes[12] because not only it is an ultimate cause of many other kinds of wastes but it prevents from identifying them. As a matter of fact, excess inventory necessitates, in Ohno's words, 1) warehouses, 2) carrying carts and 3) workers to carry goods to the warehouses; 4) workers for rust prevention and inventory management and 5) workers to repair rusted goods; 6) workers and 7) computers for inventory control.[13] That is costly though no value is added to the stocked products. As far as there is overproduction, it is very difficult to identify the other six categories of wastes and to recognize and classify them as wastes to be eliminated at all costs.

Why do companies overproduce? There are different causes to and reasons for overproduction.[14]

- Companies overproduce, i.e., manufacture more than the market needs, in order to store goods so that they can be able to respond quickly to a sudden and unforeseen surge of demand. Managing is planning and it is necessary to plan against the risk of running out of stocks that can lead to lost sales and

---

12   T. Ohno, *Toyota Production System*, pp. 54–55

13   See T. Ohno, *Toyota Production System*. 1988, p.55

14   Sh. Shingo, *Study of Toyota Production System from industrial engineering point of view*, 1981, p.87–88

customers frustration. On the other hand, a fluctuation affecting the demand in the sense of a recession leads unavoidably to unwanted goods that clearly show that overproduction is a waste.

- Companies overproduce because extra items, i.e., stocks are considered necessary to compensate for the long time of exchange of dies and tooling during which the production is stopped. The optimal quantity of such stocks is usually determined by a mathematical model known as the economic production order quantity or economic production lot size.

- Companies overproduce because stocks are considered necessary in case of production of defective or in case of breakdowns. A company that does not overproduce would be unable to satisfy its customers (internal or external) by replacing defective items with good ones, and would not even be able to meet customers' demand when a process breaks down.

- Companies overproduce in order to reduce the transportation costs relating to the work in process inventory that should be moved from one process to another or from the production process to the inspection point within a production site.

- Overproduction may occur because of the fact that machines in the production line have different production capacities. High production capacity processes will necessarily overproduce.

- Overproduction may also occur due to the difference of operation time and/ or cycle time between workers or to the difference of their skills. More skillful operators would overproduce.

As one may have realized, overproduction directly results in generating a pond of items that do not correspond to the immediately needs of the customers. The items that stagnate and make up the pond instead of flowing are called stocks or inventories. In the conventional way of thinking, stocks are assets and stocks are necessary.

In fact, for all or some of those reasons mentioned above, overproduction is felt as necessity in order to have stocks on hand. It should however be brought to the light the fact that the same conventional wisdom in the manufacturing has the merit to recognize overproduction (and the stocks it generates) as an evil. Unfortunately, it considers it as a necessary evil. For JIT, overproduction is just an evil, but the worst of all wastes, and as such it needs being eliminated completely.

According to JIT, overproduction is the cause, the source and the mother of many other kinds of waste.

Quantitative and early overproduction

Shingo distinguishes two kinds of overproduction: quantitative and early overproduction.[15]

Quantitative overproduction refers to the fact of producing more than the market needs. Producing a hundred items whereas there is only a 95 items means there is an overproduction of 5 units. As mentioned earlier, companies quantitatively overproduce in order be able to satisfy the customer demand with replacement items in case of their order would contain some defective products. Nevertheless, this remains a patent waste.

If the notion of quantitative overproduction is easily understood as a waste, early overproduction which is related to the timing or schedule of production, is hardly seen as such. Producing earlier than required is overproducing. "… *if 5000 pieces are ordered for delivery on December 20 but produced by December 15, that is early overproduction.*"[16] And this kind of overproduction seems justified by some of some of the reasons mentioned above. For JIT, it is a pure waste that must be eliminated because the stated reasons are not sound at all.

And JIT has as its main objective the elimination of all kinds of overproduction, especially the early overproduction. How does JIT cope with the problem of wastes in general and that of overproduction in particular?

JIT techniques and waste elimination

In fact, JIT which is an integral part the Toyota production system[17] has developed special techniques or methods of production that deal with each kind of wastes:

➤ *The kanban system*: it is a tool that deals with the problem of inventory control in general and that of fluctuation of demand in particular.

➤ *The quick setup* aims at reducing drastically the changeover time to its strict minimum.

➤ *The reduction of the shop floor space*, the especially U-shaped lay-out of machines/processes, the multi-machine manning system and the standard operation are responses to wastes of transportation, time on hand for workers and operators' motion.

---

15   Ibid. p. 96–98

16   Shingo, Toyota Production System, Cambrige, MA: Productivity Press, 1989, p. 69

17   There are now some variants of JIT. The present study refers mainly to the original JIT initiated by and at Toyota

➤ *The total preventive maintenance* (TPM) and the autonomation deal with the problem of breakdowns and detectives.

➤ *The quality control circles* and *suggestions systems* that draw on and tap the knowledge of all employees in order to improve the quality of the product, the production process, the machines, the workers' operations, the work environment, etc.

➤ *The autonomation techniques* that prevent machines from producing defective items, from quantitatively overproducing, from performing defective work, etc

All those techniques as constituting a whole body aim at reducing drastically the production lead time and at eliminating all kinds of wastes in the manufacturing so that the "needed items arrive at the production site at the time and in the quantity needed."[18] That body of production techniques and methods is referred to as the JIT system.

3.2. Kanban system

The purpose of the kanban system is to produce just in time. It is a tool or a means for achieving the just-in-time production, an operating method of the Toyota production system[19]. What is a kanban? There are many definitions of kanban.[20] However the following definition by Ohno seems the most satisfactory since it covers its descriptive and functional aspects:

> *Its most frequently used form is a piece of paper contained in a rectangular vinyl envelop. This paper carries information that can be divided into three categories: pick up information, transfer information and production information. The kanban carries information vertically and laterally within Toyota itself and between Toyota and the cooperating firms.[21]*

Kimura and Terada gave details on the kinds of information that a kanban usually carries:

---

18 T. Ohno, *Toyota Production System. Beyond large-scale production*, 1988, p.32

19 Ibid., p. 20, 123–124

20 *See* also R. Muramatsu "A comment on this book" in T. Ohno, *op. cit.* p. xviii; Sh. Shingo, Study of Toyota Production System from industrial engineering point of view. 1981, p.190

21 T. Ohno, *Toyota Production System. Beyond large-scale production*, 1988, p. 27

(1) *Part name, part number;*

(2) *Quantity designated usually equals a container capacity. The reorder point and ordering quantity are equal to the container size multiplied by an integer;*

(3) *Preceding process: manufacturing process, assembly line or storage location;*

(4) *Succeeding process; (as above).*[22]

Other information such as the type of packing and the number of identical kanbans issued may be also specified as a reference.

The main advantage of using a kanban is that a small piece of paper provides the needed information about the quantity, time, production sequence, the point of storage, transfer equipment, container (see exhibit of kanban types in Figures 3–5 & 3–6).

It is worth mentioning that today's a kanban features a variety of physical forms or shapes. The material a kanban is made of has no importance.

Write Nakane and Hall:

> *What is important is that a highly visible card system pinpoints control of inventory levels for each specific part, and at the same time, permits some deviation from a fixed schedule keeping the system synchronized.*[23]

The kanban must be distinguished from the kanban system. The former is an element of the latter. The kanban system should be understood as a special network of manufacturing management information system that "*that harmoniously controls the production of the necessary product in the necessary quantity at the necessary time in every process of a factory and also among companies.*"[24]

A close analysis of the definitions of the kanban and the kanban system suggests clearly that there are two kinds of kanban systems: an in-plant kanban sys-

---

22 Kimura, 0. and H. Terada, "Design and analysis of Pull System, a method of multi-stage production control." in Y. Monden, *Toyota Production System. Practical approach to production management*, Atlanta, Norcross, Georgia: Industrial Engineering and Management Press, 1983, p. 221–232

23 Nakane, J. and R. W. Hall, "Management specs for stockless production", *Harvard Business Review*, May/June 1983, p. 87

24 Y. Monden, *Toyota Production System. Practical approach to production management*, Atlanta, Norcross, Georgia: Industrial Engineering and Management Press, 1983, p. 13

tem and an inter-companies or inter-plants kanban system. The in-plant kanban system is usually referred to as in-company kanban whereas the inter-plant kanban known as the supplier's kanban system. An inter-plant kanban can be in-company or inter-company kanban. In fact, an in-company may be an interplant kanban when it is used between two plants of the same company. It works then exactly in the same way as the outside supplier and should be qualified as a supplier kanban. On the other hand, an inter-plant kanban may as a matter of fact be an inter-company kanban when the supplier plant does belong to a different company.

3.2.1. In-plant kanban system

The in-plant kanban system is made up of two complementary sheets of kanban: the withdrawal kanban and the production ordering kanban.[25]

Figure 3–5 Withdrawal kanban

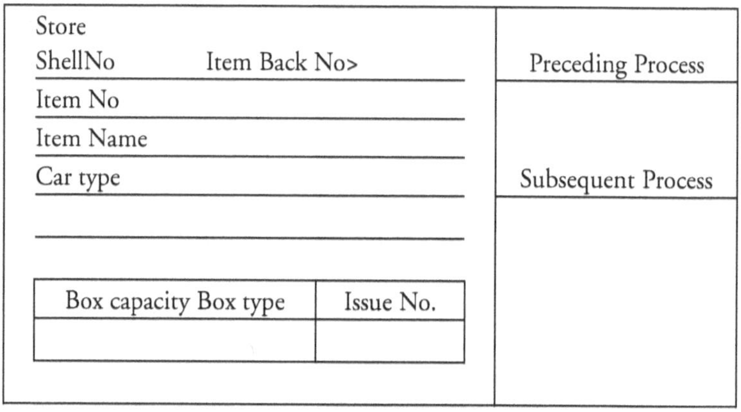

Store ShellNo Item Back No>	Preceding Process
Item No	
Item Name	
Car type	Subsequent Process
Box capacity Box type Issue No.	

Figure 3–6 Production ordering kanban

Store ShellNo Item Back No>	Process
Item No	
Item Name	
Car type	

25 Different names are used by kanban system specialists to designate the two kinds of kanban. Shingo speaks of receiving kanban and production instruction kanban; Monden, Schroer et al., Ohno: withdrawal and production instruction kanban; Nakane & Hall: move card and production card; Kimura & Terada: in-process kanban and interprocess kanban

This couple of kanbans is also referred to as the dual-card kanban by Schroer, Black & Zhailg.[26]

It is worth noting here that JIT/kanban system is a pull system, i.e. the starting point is the final assembly line. Operations initiated at the final assembly line will trigger operations backward in the entire manufacturing processes.

3.2.1.1. Working mechanism of the dual-card kanban

The main characteristic of the kanban production system is that it handles and controls small batches of products. Schroer et al. are more explicit about this point:

> *The dual-card kanban system described here has two unique features. First, carts are designed to hold small, fixed number of parts. Second, there are two kanban or cards for each cart.*[27]

Figure 3–7 shows how the dual-card kanban works.[28] It describes two stages, i.e., Processes #1 and Process #2. The latter is the supplier of parts to the former. In other words, Process #2's outputs are used by as inputs by Process #1, meaning that materials are flowing from Process #2 to Process #1 process.

(1) Functioning of the in-process kanban

While a cart full of parts is being transferred ⑤ from Process #2 storage location ⑥ to the storage point ② of Process #1, the production kanban that was initially attached to the cart been has been removed and dropped into the production kanban receiving box ⑦. The production kanbans in the kanban receiving box ⑦ will be transferred to the kanban schedule board ⑧ where they will be arranged according the production scheduling or requirement. If there is no special requirement they will be in the same order they were put into the kanban box. The order in the kanban schedule board indicates the sequence in which different items will be processed or produced at the work cell: in the mix production system, a work cell may be producing different kinds of items. When a standard container is filled

---

26 Schroer, B. J., J. T. Black and Shou Xiang Zhang, "Microcomputer analyzes 2-card kanban system for 'just-in-time small batch production", *Industrial Engineering,* June 1984, pp. 54–64

27 Ibid. p. 55

28 I tried to adapt Schroer, B. J., J. T. Black and Shou Xiang Zhang, (1984) and Kimura, 0. and H. Terada, (1983)

with parts according to the instruction of the kanban, that the kanban is attached to the container and the full cart is stored at the stock location ⑥ of Process #2.

(2) Broadening the true meaning of kanban

As described above, the kanban consisted originally of two pieces of paper conveying the production and transfer order. Let us once more remind the fact that the most important is not the materials a kanban is made of, but the message and its timing as a means of controlling the production and the movement produced and supplied items within the production site. A cart, for example, depending on the fact that it is empty or cart full of items, may play the role of the production kanban or the transfer kanban respectively. A specifically designed and reserved space on the production floor, when empty, may be a sign to start the production whereas a container occupying the same space might mean, "Do not produce but transfer the container to the subsequent process". Of course, electronic devices such as board displayers, computer monitors, etc may also convey production and withdrawal information.

(3) Functioning of the inter-process kanban

When the production work cell ① of Process #1 starts using items from a cart at the storage location ②, the transfer kanban is removed from the cart and put into the transfer kanban box ③. The kanban should be removed since the content of the cart does not any more match the number of items written on the move kanban.

At regular intervals, a worker would bring those kanbans from the box, and take them with empty containers the storage ⑥ of Process #2. The empty containers will be left at the storage location. The kanban will be attached to the container full of products. Before attaching the transfer kanban to the container full of products, its production ordering kanban is removed and put into the in-process kanban box ⑦. The cart full of parts with its transfer kanban can then be moved ⑤ to Process #1 storage place.

Figure 3–7 Working mechanism of the dual kanban

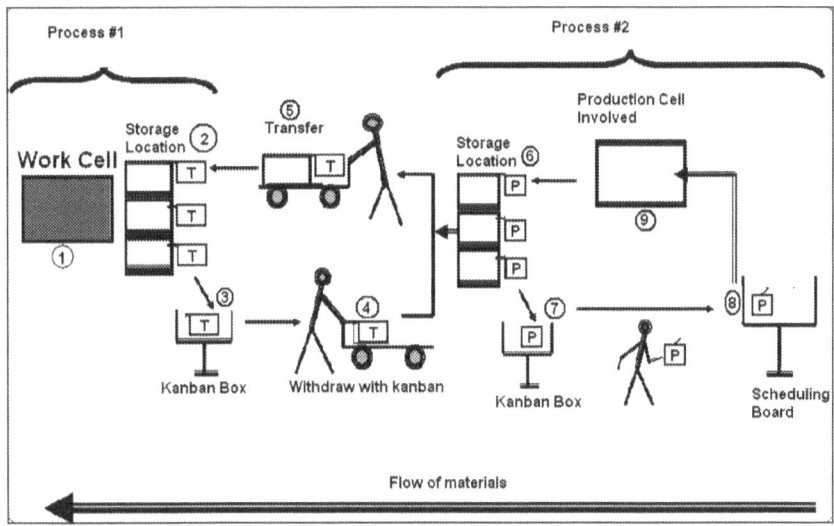

3.2.1.2 Rules of the dual-card kanban

The working of the kanban system obeys strict rules[29] without which the kanban would be of no use at all. The rules may be classified into three categories: common rules regulating both kanbans and rules specific to each of the two kinds of kanban. To these basic rules will be added supplementary rules that concern defective items.

a) Common rules:

1. A kanban should be always attached to any container filled with the required number of products or parts.

2. The number of kanban should be kept as small as possible.

b) Inter-process kanban rules:

3. The quantity of products to be withdrawn should be equal to the quantity indicated by the inter-process kanban.

4. Any withdrawal without the authorization of a withdrawal kanban is prohibited

---

29 T. Ohno, *Toyota Production System. Beyond large-scale production,* 1988 p.30; Y. Monden, *Toyota Production System. Practical approach to production management,* 1983, p. 23–28;Sh. Shingo, *Study of Toyota Production System from industrial engineering point of view,* .1981 p. 292

c) In-process kanban rules:

5. The production of the supplying process should correspond to the quantity withdrawn as indicated by detached production kanban.

6. No items should be produced without the authorization of an in-process.

7. When various kinds of parts are to be produced in the preceding process, their production should follow either the original sequence in which each kind of kanban has been delivered or the sequence of kanbans as indicated on the schedule board.

d) Additional rules

8. No defective parts or items should be forwarded to the succeeding process.

9. A received defective part or item once detected should be immediately returned back to the preceding process that made it.

The first seven basic rules may be condensed and reduced then only to three as did Nakane and Hall[30]:

1. *Always use standard container filled with the correct number of parts.*

2. *Never move a standard container forward without getting the authorization of a move card.*

3. *Never produce a standard container of parts without the authorization of a production card.*

Schonberger states also a three-rule set for the dual kanban but in a different manner:

1. *No parts may be made unless there is a production kanban authorizing it*

2. *There is precisely one conveyance and one production kanban for each container.*

3. *Only standard containers may be used and they are always filled with the prescribed (small) quantity—no more, no less.*[31]

---

30 Nakane, J. and R. W. Hall, "Management specs for stockless production", *Harvard Business Review*, 1983, p.85

31 R. J. Schonberger, *Japanese manufacturing techniques. Nine Hidden lessons in simplicity*, N. Y.: The Free Press, 1982, p. 224

3.2.1.3. Role or function of kanban and kanban system

A kanban fulfills the function of an operations instruction ticket and that of an identifying tag.[32]

The identification function is common to both in-process and inter-process kanban. But the instruction type depends on the kind of kanban: it is an operation instruction for the in-process kanban that tries to answer the following question: what product, till when and in what quantity should it be made? For the inter-process kanban, the instructions concern the transfer of products between two processes: from where to where should the container filled with the prescribed quantity of parts be moved? The main function of the kanban system is to control inventory, regulate stocks and adapt the production to meet the deviation of the actual demand from a projection or a demand forecast. In the case of a demand increase, the circulation speed of kanban and/or their number will be increased. When demands decrease, the movement of kanban or their number will be decreased.

One of the consequences of using the kanban system is that, by always striving to reduce the number of kanban to the minimum (see kanban rule), the number of items flowing in the assembly line is limited thus to the necessary minimum.

Keeping the stock at the necessary minimum is already a solution to the worst waste of overproduction, and by having only a strict minimum of stocks, any production problem that any arise (for instance, occurrence of defective parts) is immediately brought to the light and thus calls for immediate remedial actions.

In a word, the function of the kanban system is to prevent overproduction and excess transportation, to reveal existing problems and to control inventory. As for the dual-card kanban, its main function consists in providing pickup, transportation and production instructions.

3.2.2. Supplier kanban system

The descriptive definition of a kanban borrowed from Ohno at the beginning of this section mentions among other things that "kanban carries information vertically and laterally within Toyota itself and between Toyota and the cooperating firms". The cooperating firms are to be understood as suppliers and the kanban that connects Toyota and a supplier is referred to as a supplier kanban.

Mentioned below are some elements of the supplier kanban system: (a) a supplier kanban to which corresponds (b) a container, (c) three trucks, (d) three people: a driver and two workers involved in loading and unloading operations.

---

32 See Sh. Shingo, *Study of Toyota Production System 1981*, pp. 272–273, 281–283

Whereas the driver goes from and to a supplier's plant, one worker is loading at the supplier's plant an empty truck with containers filled with manufactured parts while the second worker, at Toyota, is busy unloading the third truck full of supplier's previously delivered parts.

The driver will leave at the supplier's plant the truck of empty containers and the supplier kanban (in case of hard copy kanban) and in driving to Toyota the truck already loaded with full boxes. At Toyota, he will leave the truck full of parts from the supplier and get into the truck already loaded with empty boxes and head to a supplier's plant to pull parts from the supplier to Toyota. There would thus be no idle time for the truck driver.

The working mechanism of the supplier kanban will be grasped first at the supplier's plant and second at Toyota.

3.2.3. Supplier kanban at the supplier's plant

In general, the supplier starts producing parts when he receives the supplier kanban that the driver will bring to him at the withdrawal time for those parts already loaded in the awaiting truck. Therefore, the supplier kanban plays a double role: it fulfills the function of transmitting both the production and transfer instructions. Iizuka and Monden explain this double role:

> Supplier kanban used as a common kanban for dual purposes (production and supply) initiates parts production processes because it functions as production instructions as well. At the time of withdrawal of parts manufactured previously for delivery to the manufacturer, a new supplier kanban is brought in to the supplier's production process, and the new supplier kanban assumes the function of production kanban. The same kanban resumes the function of supplier kanban when production of parts is completed and the time of the delivery of the parts to the manufacturer. Thus, a single kanban has two functions.[33]

3.2.3. Supplier kanban at Toyota (JIT case)

The containers of parts manufactured by the supplier are stored at a special point and brought to the work cell when needed and when called by a signal light called "andon". According to Monden,[34] when at a work cell, a box becomes empty,

---

33 Iizuka, Y. and Y. Monden, 'Mechanism of supplier's response to the kanban system." in Y Monden, (ed.), *Applying just in time: the American/Japanese Experience*, Norcross, Georgia: Industrial Engineering and Management Press, 1986, p.77

34 Y. Monden, *Toyota Production System*, 1983, p. 52–53

the worker will push a switch that will light on a lamp at the storage point and on seeing the light, the carrying worker will detach the kanban, put it into a "post office" for supplier kanban and then bring the box of materials to the work cell that needs to use them.

The detached kanban will be classified for each supplier, and later given to the truck driver who will take them to the different suppliers.

At the storage point, the supplier kanban attached to the full box may be considered a production-ordering kanban and the "andon" light could be viewed as withdrawal instructions (kanban). The two kanbans play the same roles as the dual-card kanban described earlier, obey the same rules and fulfill the same functions. The main different is that the preceding process is here the supplier plant.

3.2.5. Kanban system, market research and production smoothing

JIT does not allow "speculation production" on the one hand and on the other it is so delicate that it can hardly handle big and sudden fluctuations of demand. That is why the kanban system at Toyota is supported by a strong market research.

Without getting into details, let me recall the fact than, at Toyota, there is a yearly production plan. But the working production plan is a three month production schedule:

> *Toyota sends a three-month production schedule, called a parts requirement forecast table, to its parts suppliers. Information about actual parts supplied during the most recent month is provided as a final forecast and recorded on a daily basis. The forecast for the remaining two months is estimated.*[35]

The daily plan derives from the monthly plan which is broken into three periods of ten days. The ten-day plan consists of placed orders by Toyota dealers. In fact, the kanban system as practiced by Toyota produces only salable products i.e., like in a super-market that replenishes its shelves, JIT-based manufacturing produces to replace sold items.

At this point it seems necessary to introduce and understand the concept of production smoothing.

Putting aside technical explanations, the smoothed or leveled production is indicated by the daily sequence production. It is the averaged production per day for all the types of produced items in general and for each type in particular. The sequence is determined according to sophisticated computer methods; then it will

---

35   Y. Monden, *Toyota Production System*, 1998, p. 79

be notified only to the final assembly line by a printer or a display apparatus of a computer terminal[36]

3.2.6. Recapitulation

It was stated earlier that the kanban system should be viewed as a special manufacturing management information system. It conveys the manufacturing information backward, the starting point being the final assembly line.

The system links not only the production cells of the final assembly line but also it connects the latter to supplier companies.

The main feature of the system is the kanban, of course. It is a signal element authorizing the production of items or their transfer in the prescribed quantity, at the time required time and where they are needed for use or storage. The work-in-process inventory is thus limited by the number of kanban being used. As it is required that the number of the kanban be reduced to its necessary minimum, the inventory of work in process is therefore reduced to tie minimum (over-production of work in process is thus partially eliminated.)

It should also be brought to the light the fact the kanban system is a pull system. That is why the (final assembly line is the starting point. And at this starting point the kanban that authorizes the production or the transfer of parts to be used is a reaction, an answer to accepted order of a consumer. That means that the kanban system authorizes only the production to meet the actual demand of the market (over-production of finished goods is thus partially eliminated).

However, it is impossible to produce just in time so that the demand of the moment may be met without any previous knowledge, be it vague, about the approximate number of the items to produce. This knowledge, though not so accurate, helps prepare roughly the number of people and the approximate quantity of materials that the company may be required to use. This number should be considered provisional and thus subject to modification to meet the needs of the real demand. Therefore, the importance of the market research and production plans or projections is brought to the light since the kanban is almost unable to handle great and unforeseen fluctuations in demand. The kanban replenishes only what has been planned.

The attention should be drawn to the following element: in the conventional manufacturing system, it always takes many hours to exchange punching press dies weighing many tons. Accordingly, the conventional wisdom requires that at each production run, each part be made in large quantity (large lot size or mass production) before stopping the machine for dies changeover. But it is clear that

---

36   Ibid. p.59–60; see also Ch. J.McMillan, *The Japanese industrial system*, Berlin: Walter de Gruvter, 1985 (second revised edition), p.2 14

the "mass production of a single item", be it the optimal or economic lot size, can hardly get close to or even never match the real needs of the market because different people do not decide, say to order the same kind of cars at the time and the other kinds at another time. Each consumer places his/her order according to his/her tastes, preferences, needs, financial means, etc. Therefore, the mass production of one item at a time leads unavoidably to stocks, i.e. over-production that JIT considers the worst of the wastes.

To produce according to the actual demand of the market requirements, the kanban-supported JIT has to overcome first the problem of the die change-over. And JIT as initiated at and by Toyota has succeeded in reducing the changeover time.

3.3. Necessity of reducing setup time

The kanban system is without any doubt not only the most obvious and specific feature of the JIT production but also one of its most efficient tools. Yet, the kanban system itself which allows only small but frequent lot production deliveries is almost impossible to apply unless set-up time has been drastically reduced. What I intend to say is that JIT production is conditioned by the kanban system and the latter is conditioned by a short time of changeover. The shorter that time, the better the system. Clegg states clearly the importance of a quick set-up:

> *One of the most important step a manufacturing facility should take before considering implementing just-in-time inventory control is reducing set-up time to the lowest practical value.*[37]

The reduction of the set-up time as a prerequisite for the kanban system will be dealt with in four points: 1) the rationale of the economic order quantity (EOQ), 2) its rejection by JIT and 3) the effects of the set-up time reduction on the production lead time.

3.3.1. Economic order quantity/economic lot size

The rationale of the economic order quantity or economic lot size (EOQ|/ELS) can be grasped using the Hegelian dialectical method.

Hegel's dialectical reasoning method, as everybody knows, is made up of three steps: (1) the thesis states something; (2) the antithesis denies it; (3) the contradic-

---

37 Clegg, W. H., "Operator/machine studies Technique reduces set-up time, implements JIT", *Industrial Engineering*, October 1986, p.52

tion created by the thesis and the antithesis is solved by their synthesis that is an instance of reconciliation at a higher level.

Table 3–2 Comparison of changeover times of stamping machines by country (hood and fender)

	Toyota	A. Co. (USA)	B. Co.	C. Co. (W. Germany)
Stop of stamping machine during exchange of dies	9 min	6 hr	4hr	4hr
Number of exchange	1.5/shift	1 in less than 2 shifts	-	1 in 2 days
Size of lot	1 day	10 days	1 month	-
Stroes per hour	500–550	300	-	-

Source: Shingo, *Toyota Production System*, 1981 p. 169; 1989, p. 107

Let me apply these elements to the production system.

(a) <u>Thesis</u>: Reduce set-up costs by using the large lot production system. In the conventional manufacturing, machine set-ups take many hours (see Table 3–2). So to reduce the setup costs, it seems logical to produce the same kind of item in large quantity before changing to the production of another kind. The immediate consequence is the large lot production is the piling-up of inventory of both wip and finished products, resulting in the increase of inventory holding costs.

(b) <u>Antithesis</u>: Reduce the inventory carrying costs by small quantity production. In fact, producing large quantities means that inventory level and the costs relating to managing and carrying it will go high. Schonberger reminds of what carrying charges are:

Carrying charges are the interest costs on capital tied up in inventory, plus the physical holding costs, such as warehouse rent and warehouse workers wages.[38]

Therefore in order to reduce the carrying charges for inventory, it seems quite logical to produce frequent but small quantities of different items. Of course the ideal quantity would be just one unit.

(c) <u>Synthesis</u>: EOQ/ELS

(a) and (b), though aiming both at the same goal, i.e., cost reduction, destroy each other. However if (a) and (b) are taken as the premises of a dialectical reasoning, the so-called EOQ stands as their conclusion i.e., the instance of their reconciliation.

---

38 R. J. Schonberger, *Japanese manufacturing techniques*, 1982, p.18

In fact, in order to reduce both the setup costs (thesis) and the cost of carrying inventories (antithesis), a compromise quantity that is not too big or too small has been mathematically defined and is well known as EOQ or ELS.

Since the conventional manufacturing systems are run according to this dialectical reasoning, one understands why they do have so much inventories of work in process or of finished goods.

Hayes states reasons why US firms run large inventories:

> *Why do US companies have such work-in-process inventories? One major reason is their emphasis on producing economic batches which seek to balance inventory costs against the set up costs created by changing from one item to another.*[39]

3.3.2. JIT rejects both EOQ and its rationale

Yet, JIT-based manufacturing does not accept the EOQ. JIT considers it not only as a pure waste but as the worst of all kinds of wastes, i.e., over-production. This system rejects EOQ and the rationale that underlying it and justifying its necessity. As a mathematical optimization model, it is quite fine. It was very useful in a mass consumption market, but not in the highly competitively customized markets of today.

Instead, JIT reasoning goes as follows: Why does the overproduction caused by EOQ have to be accepted as a necessary evil? It is because changeovers take many hours. Why doe setups have to take so long? Why not try to shorten the changeover time?

Conventional manufacturing systems consider a long set-up to be a given, a parameter of which the value (time) cannot be changed and to which is tied a fixed cost.

For JIT, this point is however questionable. JIT companies have succeeded in reducing setup times from hours to minutes (see Table 3–2). As recently as 1992, at Johnson Controls, an America company and seats supplier to Toyota, the setup time was 6 hours. Toyota worked with that company to reduce it to a mere 17 minutes[40]

And, when the set-up time is minimized, the logic of EOQ loses its basis and looks obsolete. Johansen and McGuire put it this way:

---

39 Hayes R. H., "Why Japanese factories work", *Harvard Business Review,* July/August, 1981, p. 59

40 J; K. Liker, "Japanese automakers, US suppliers and supply chain superiority", **Sloan Management Review**, Fall 2002, p. 83

When set-up times are measured in minutes instead of hours, the logic of the economic order quantity (EOQ) approach is dissolved. The blind-spot of EOQ logic has been that set up cost are both fixed and high.[41]

The ideal for JIT is to reduce to zero the time of changeover i.e., a complete elimination of setup costs.[42]

Ohno emphasized better than I could the importance of shortening the changeover time:

Only if we succeeded in minimizing the time where it was negligible, could lot quantity be decreased and running stocks also be diminished. Ideal just-in-time production is only a dream unless die set-up time is reduced. We were finally, through great effort, able to shorten our die set up time within minute.[43]

And he goes on further stating:

Now we are studying exchange of dies and tools within second which is easy to express but actually a very difficult problem. But anyway, we have to shorten the time of exchange of dies and tools.[44]

3.3.3. Effects of a quick set-up on the production lead time

It is worth note that for any given process the reduction of setup time implies an equivalent shortening of the production lead time. What I intend to say is that for any work cell or processing unity,

(1) production time = processing time + setup time

To illustrate the statement, I am going to examine the following example (a theoretical, hypothetical and oversimplified case). Take the mix production of two items, say A and B by a stamping machine.

---

41   Johansen, P. and K. J. Mc Guire, "A lesson in SMED with Shigeo Shingo", *Industrial Engineering,* October 1986, p.26

42   An ideal setup no setup = zero time setup see Abegglen, J. C. & C. Stalk, Jr., *Kaisha, the Japanese corporation,* 1987 and Sh. Shingo, *Study of Toyota Production System from industrial engineering point of view,* 1981

43   T. Ohno, "The origin of Toyota Production System and Kanban system. in Monden, (ed.), *Applying just in time: the American/Japanese Experience,* Norcross, Georgia: Industrial Engineering and Management Press, 1986, p.6

44   T. Ohno quoted by Sh. Shingo, *Study of Toyota Production,* 1981, p.70

To switch the production from one item to another, four hours are needed for dies' changeover (as was the case at Toyota before 1970 and as is the case in some conventional manufacturing systems today). Suppose the processing for making each item requires only 5 seconds:

(2) processing time per unit = 5 seconds

If one wants to produce both items, say, one A and one B, making one A will take five seconds; then four hours will be spent on changeover operations and five other seconds will be required for stamping one B.

Therefore, 4 hours 10 seconds will be needed to make these two items, say one A and one B alternatively:

(3) production lead time of A & B = 5 sec. + 4 hr + 5 sec. = 4hr10 sec

That means that half a day of work would be necessary to produce two items whose real processing time is only ten seconds.

However, if the changeover time is reduced to some nine minutes (as is the case at Toyota), the production of both A and B alternatively would require only 9 minutes 10 seconds.

(4) JIT lead time of A & B = 5 sec + 9 min. + 5 sec = 9 min 10 sec

The lesson and message for manufacturers conveyed by the example are that a quick setup reduces the processing cost, the labor cost and the production lead time. On top of all this, it can free the manufacturer from the need of storing products.

3.4. Methods for reducing lead time

As everybody knows, storage, transportation, waiting time for workers and unnecessary movement/motion of the workers are labeled by JIT as wastes. They increase the production lead time, hindering thus the just-in-time production. As a second and direct effect, they increase the production costs. How does JIT deal with those wastes? JIT considers that those problems are related to the shop floor space, the machinery layout, the relationship workers and machines/processes, the workers' operations and the number of workers on the shop floor.

3.4.1. Shop floor space

In the plants using the conventional manufacturing system, plant floors are so 'spacious', borders between various processes so 'clear' and so 'sharp' that even the setup time is reduced to its minimal level, it remains always necessary to store produced items not only for transportation reasons but also because it seems necessary to do so before these items can cross the borders separating different processes.

For JIT, the storage of products for any reason is to be absolutely avoided. But when the shop floor is very large, the distances and the sharp compartmentalization between processes become causes of storage. As a matter of fact, in the large shop floor, if small quantities/lots of items are produced, the transportation frequency and the cost related to it will unavoidably increase. And transportation is a pure waste which not only raises the costs but also hides, as Ford noticed it many decades ago, the growth of a company:

> *The real limit to the size of a corporation is transportation. If it has to transport its commodity too far, then it can not give service—and it limits its own size.*[45]

Shingo echoes Ford's concerns but he is tougher about transportation as a waste:

> *The phenomenon of transportation is not to increase added works but only elevate cost performance. (…) even if manual transport work is mechanized, it simply means that requirement of high transport work cost was converted from manual to mechanical which is a loss without returns. Consequently, we must maintain a strong attitude considering absolute elimination of transportation.*[46]

JIT has discovered that wastes of transportation and those of storage related to the latter occur due the long and/or complicated distances between processes.[47]

Therefore, the only solution to the problems of transportation and storage is the reduction of the shop floor space. The expected result is as described in the following passage by Hill:

> *As the overall size of the plant is reduced and the average lot size is also cut, changes will occur in the need for internal transportation. Long and complex conveyor runs can be eliminated, numbers and capacity of forklift*

45 Ford, H., *Today and tomorrow*, Cambridge, Mass.: Productivity Press, 1988 (Reprint. Originally published: Garden City, NY: Doubleday, Page & Company, 1926), p.20

46 Sh. Shingo, *Study of Toyota Production System*, 1981, p.38

47 See R. J. Schonberger, *World class manufacturing: the lessons of simplicity applied*, N.Y.: The Free press, 1986, p. 86

trucks can be reduced, need for storage racking is removed and precision of positioning of materials becomes more important.[48]

JIT companies try to reduce the shop floor space as much as possible. The space may be reduced by more than 70% as is the case of CSR Division of Cleveland Machine Controls Inc. which, according to Kachur,[49] switched to just-in-time manufacturing and then implemented an overall reduction of about 73%.

3.4.2. Borders between processes

Although reducing the plant floor space contributes effectively to shortening the transportation time between processes, the phenomenon of transportation is still there. How to eliminate completely the waste of transportation? At least, how different processes should be laid out so that the problem of transportation can be minimized? Many specialists of the JIT production are unanimous on the following three principles concerning the most efficient configuration of machines layout.[50]

Principle-1: Break down barriers between departments or processes and layout machines as closely as possible. As barriers between processes/working cells disappear, mutual assistance becomes possible (See section 6.2)

---

48 Hill, I.D., "Modern manufacturing techniques require flexible approach to facilities planning", *Industrial Engineering*, May 1984, P. 86

49 See Kachur, R. G., "Electronics firm combines plan move with switch to JIT manufacturing", Industrial Engineering, March 1989, pp. 44–48

50 See J. J Feather and K.F. Cross, "Workflow analysis. Just-in-time techniques simply administrative process in paper work operation", *Industrial Engineering*, January 1988, pp. 32–40; Hill, 1.1)., "Modern manufacturing techniques require flexible approach to facilities planning", Industrial Engineering, May I 984, p.88; Y. Monden, *Toyota Production System,*1983, pp. 68–74; 99- 116; R. J. Schonberger, "Plant layout becomes product-oriented with cellular, just-in- time production concepts", *Industrial Engineering*, November, 1983, p.71; She. Shingo, *Study of Toyota Production,* 1981 .36–38, 153–167, 288, 304-SOS; J. A. Tompkins, "Successful facilities planner must fulfill role of integrator in the automated environment", Industrial Engineering, May 1984, p.54–58

Figure 3–8 Production lead time at a one-at-a-time processing line

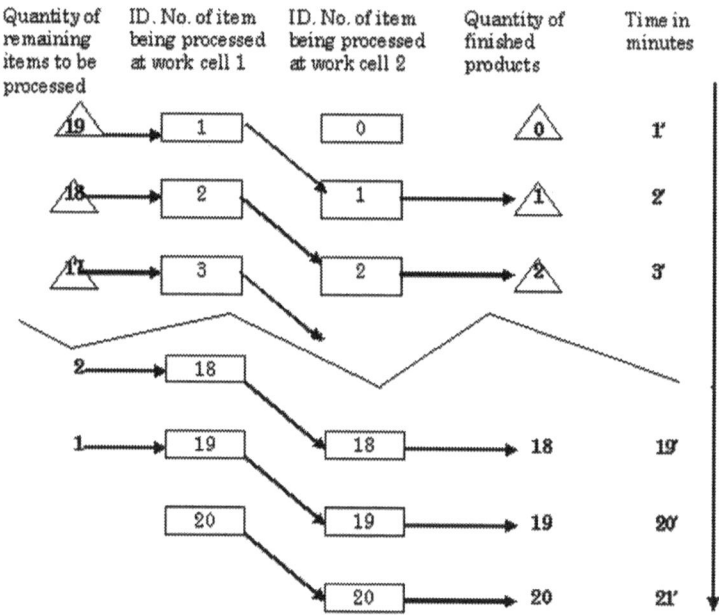

Quantity of remaining items to be processed	ID. No. of item being processed at work cell 1	ID. No. of item being processed at work cell 2	Quantity of finished products	Time in minutes

Principle-2: Integrate various manufacturing processes into a 'single machining center'[51] so that 'one-at-a-time processing'[52] should be realized. It means that sustained efforts should be made in order to create a single unit production and conveyance.

To create a one-at-a-time processing line requires that different kinds of machines be arranged in accordance with the flow of processes' products instead of by type of machines. Once a continuous flow is realized, the wip storage and transportation between processes as well as the production lead time will be reduced.

Let me illustrate the reduction of the production lead time by the following example. Assume that there are two lots of 10 units each that must go through many processes of transformation and that it takes a single minute for processing one unit. Let us together observe closely the first two processes only. Figure 3–8 shows that the first will be completed being processed after two minutes. Then,

---

51 A. Tompkins, op. cit. p. 54

52 Feather J.J. and K.F. Cross, "Workflow analysis. just-in-time techniques simplify administrative process in paper work operation", p.36

the throughput will be one unit per minute. In other words, it will take two minutes to process the first unit and 19 minutes to do the 19 other items.

The processing time may be computed mathematically. In the case the production run quantity, **Q**, is processed entirely at each process, and it takes **t** units of time to process an item at each process, and if there are **P** processes on the other hand, then,

(1) *Processing completion time* = **PQt**

If the production run quantity **Q** is divided in smaller lots of equal size, **q**, then the lot processing time per lot per process can be computed as follows:

(2) *Lot processing time/process* = **qt**

The number of lots, **L**, is of course the quotient of Q and q:

(3) Number of lots L = Q/q, and

(4) *Processing time of Q/process* = *(qt)L* = *Qt*.

If there are more than one process, then

(5) *Total processing completion time* = *qt { P + (L-1) }*

And integrating (3) and (5), that is replacing L by its equivalent value given (2), we get:

(6) *Total processing completion time* = *qt { P + ((Q/q) − 1) }*
$$= (Pq + Q - q) \, t, \text{ or}$$
$$= \{ (P\text{-}1) \, q + Q \} \, t$$
$$= Qt + (P\text{-}1)qt$$

Example

Forty items are to be produced and it takes 2 minutes for an item to be processed at each of the five processes of our plant. Find the total production time if items are to be processed in lots of 40, 20, 10, 8, 7, 5, 4, 3, 2 and 1 item.

P = 5, t = 2 min/item, Q = 40 items

- If q = 40, then
  Production time = ((5-1)40 + 40)2 = (4(40)+40)2 = (40(4+1))2 = 40(5)(2) = 400 minutes
- If q = 20, then
  Production time = ((4)20 + 40)2 = 240 minutes
- For q = 10, 8, 7, 5, 4, and 3 the production times are 160,144, 136, 120, 112 and 104 minutes respectively
- If q = 2, then
  Production time = ((4(2) + 40)2 = 96 minutes
- If q = 1, then
  Prod time = ((4(1) + 40)2 = 88 minutes

As one can see, the formula (6) is an expression of the total processing time in terms of (P, t), **Q** and **q**. The first term of the formula, Qt, indicates the processing time at process one whereas the second term, (P-1)qt shows the processing time of one lot at all the remaining processes. In its simplest form, it is the processing time of all item, Q, at one process plus the processing time of one lot at all other processes, (P-1).

The processing time can be also expressed in terms of **L and Q**. In fact, by substituting q by Q/L in formula (6), we get,

(7) *Total processing completion time = Qt + (P–1) t Q/L*
$$= (1 + (P-1)/L) \, Qt$$
$$= (L + P-1) \, Q \, t/L$$
$$= Qt(L + P-1)/L$$

The substitution of Q in (7) or in (6) by Q = Lq yields the expression of processing time in terms of **L and q.**

(8) *Total processing completion time = Lqt ((L + P– 1)/L (from 7)*
$$= qt \, (L + P-1), \, or$$
$$= Lqt + (P-1)qt \, (from \, 6)$$
$$= qt \, (L + P-1)$$

In formulas (7) and (8), the second factor, L+P-1, can be thought as expressing the number of processes through which the lot will be processed or the number of lots that should processed at one process. Formulas (7) and (8) state that the total processing time is a product of that number and the processing time of one lot (qt or Qt/L).

If Q = q then, L= 1, and

(1) *Total processing completion time = Qt + (P-1)qt = Qt + PQt – Qt*
$$= PQt, \, or$$
$$= (L+P-1)qt = (1+P-1)Qt = PQt$$

3.4.3. U-shaped processing lines and multi-machine manning system

<u>Principle 3:</u> the processing line should be U-shaped. The U-shaped layout of processing lines as initiated at Toyota has become a specific feature of any JIT company. That configuration of machinery features a high flexibility in dealing with the problem of increasing or decreasing the number of workers on the processing line. Another advantage of a U-shaped line over a linear line is the fact that, for the former, input and output are usually at the same position and they may be under the supervision of the same worker.

Besides, U-formed lines may be combined into one integrated line and such a combination adds to work force flexibility on the processing line.

In the conventional manufacturing system, each machine is manned at least by one worker. And to assure the continuous flow of products, the linear layout prevails. That is the case for example of the Ford-style production system. The underlying idea is that the worker should not move but that the work should be brought to the worker. Ford states the principle this way:

> *The thing is to keep everything in motion and take the work to the man and not the man to the work. That is the real principle of our production and conveyors are only of many means to an end.*[53]

In JIT plants, the number of machines far exceeds that of workers and each worker is busy moving from one machine to another. The JIT objective is to reduce the cost by eliminating the wastes and if each worker handles only one machine, the idle time, i.e., the waste of time on hand may occur. Many decades ago Ford noticed that the waste of time is very insidious:

> *Time waste differs from material waste in that there can be no salvage. The easiest waste of all wastes and the hardest to correct is the waste of time, because wasted time does not litter the floor like wasted material.*[54]

In order to eliminate the waste of waiting time, JIT requires that a worker performs many operations, mans as many machines as possible. And to reduce the movement of the worker (another waste) to its minimum, the U-form layout seems the most suitable.

To get an idea concerning the number of machines a worker can handle in a JIT company, Shingo reported that, at Toyota, in the 1940s and 1950s, 3500 sets of machines were manned by 700 workers, i.e., an average of five machines per worker.[55] In the 1980s, a worker could attend up to 16 machines.[56] Such a worker is referred to as multi-function worker or multi-machine manning worker.

In a word, it can be remarked that JIT accepts the waiting time of machines but not for workers.

---

53   Ford, H., *Today and tomorrow.* 1988, p. 103

54   Ibid. p.114

55   Y. Monden, *Toyota Production System,* 1983, p. 82

56   Ibid. p.69–70

JIT thinking is that the man-hour is more costly that the machine-hour, at the long run. So the number of the workers has to be reduced as much as possible since it is better for machine than for man to be idle.

3.4.4. Job rotation and OJT

It is clear that for a worker to become a multi-machine operator, it is necessary for him to undergo a special training at the job site.

Japanese JIT workers undergo the necessary training through what is well known in Japan as job rotations and on-the-job training (OJT) system.

3.4.5. Standard operations

Job rotation programs are difficult to carry out at the work site if there are no clear written procedures for the trainer and the trainee to follow. Instructions about the work procedures are so clear at Toyota that only three days suffice to get a worker master a new job.[57]

To attain the objective of three-day training, JIT-companies like Toyota, establish standard operations. A standard operation procedure features three elements: cycle time or tack, operation routine and standard inventory.

3.4.5.1. Cycle Time

The cycle time is the time span during which one unit of a product must be produced. The cycle time is computed by dividing the daily working hours by the required quantity to produce per day. Cycle times at different processes are synchronized so that when a worker finishes his work, he will pass it directly to the next worker for the next processing step.

According to the fact that the cycle time is long or short, the number of machines to be handled by one worker will be proportionally increased or decreased and inversely the number of workers on the production line will be accordingly reduced or increased.

3.4.5.2. Operation routine

Operation routine or operation sequence refers to the sequence or the order of operations in which a multi-machine worker handles items and machines. It is an instruction sheet showing the worker the orderly sequence of successive operations at each machine within the cycle time.

---

57   T. Ohno, *Toyota Production System* 1988, p.22; Sh. Shingo, *Study of Toyota Production System,* 1981, p.130; p; 221; p. 224

3.4.5.3. Standard inventory

Standard inventory refers to the minimum Work-in-process inventory between processes. This includes also the items mounted on machines.

The ideal standard inventory is made up of the only items being processed. That happens if and only if there is a one-piece flow of materials.

The standard operation sheet is displayed where everyone may see it easily and unavoidably. As such, it plays the role of a visual control. In fact, not only it is the worker's guideline but it facilitates the work of the supervisor. At a glance, the latter can see whether the movements/motions of a worker are in accordance with the standard operation instruction sheet or not.

As it can be seen, JIT really emphasizes not only the reduction of workers but also that of inventory. However, doesn't the JIT factory need some inventory stock to cope with the problems of breakdowns? How to avoid overproduction in the production line involving few workers and automatic machines of high and low production capacity? Autonomation is the technique JIT uses to deal with such problems.

3.5 Autonomation

What does autonomation mean? What is it? The concept of autonomation applies to the worker as well as to the machine. What is an autonomous worker or autonomous machine? With regard to human factor, autonomation refers to the decision power and the responsibility of every line worker to stop the entire production line if he deems it necessary, i.e., in case of trouble occurrence. In the conventional manufacturing setting, the decision to stop the line is of the competence sphere of the management.

In a JIT factory, stopping the line can be done not only by a man but also by a machine. Autonomation refers mainly to an autonomous action or decision made by a machine to shut down itself or a production line. Ohno described such a machine in the following terms:

> At Toyota, a machine automated with a human touch is one that is attached to an automatic stopping device. In all Toyota plants, most machines, new or old, are equipped with such devices as well as various safety devices, fixed-position stopping, the full-work system and baka yoke fool-proofing system to prevent defective products. (…) In this ways, human Intelligence or human touch is given to the machine.[58]

---

58 T. Ohno, *Toyota Production System, 1988*, p. 6

When do machines shut down autonomously? Autonomous machines do so for three reasons:

a) to prevent a worker, who can not or does not finish his work within the cycle time from entering the next process;

b) to avoid overproduction;

c) to avoid defective production.

And JIT companies have especially imagined various devices which stop the line or the machine for each of the mentioned reasons

3.5.1. Prescribed-position stopping method.

This method is made up of devices that stop the line automatically when the work is not done within the required time. In such cases the line should be stopped, otherwise the situation will lead to overproduction of work in process inventories at other processing stations.

3.5.2. Full-work system

The purpose of the so-called be full-work system is to prevent high production capacity machines from overproducing. In fact, on the processing line, automated machines may have different production capacities. In order to keep the standard inventory constant, high capacity and low capacity machines should be coupled and harmony should be established between them so that high capacity machines follow the production rhythm of the low capacity machines. In order words, high capacity machines should run intermittently.

3.5.3. Poka yoke and defect prevention

Poka yoke is an inspection method. The main concern of the inspection is to determine the acceptability of products as good or their rejection as bad. In other words, the object of inspection is the quality of a product.

In the conventional manufacturing systems, where large lot production is the common practice, only a sample is inspected. The reason is that inspecting each product seems too demanding and very costly. The following excerpt from the speech of an executive of a non-JIT company is a good illustration.

> *100% inspection is troublesome, so appropriate inspection assured by statistical random judgment is far more ideal. I myself believe that undoubtedly,*

sampling inspection has an assurance of statistical random judgment and much better than 100% inspection.[59]

According to Buffa & Sarin,[60] the sampling inspection accepts a lot when it contains 2% or less of defectives. This percentage determines what is referred to as the acceptable quality level (AQL) Reddy & Berger express the idea on which AQL notion relies:

The AQL approach assumes that a certain small percentage of defective products is inevitable and that the object of quality control is to reach the AQL level.[61]

JIT rejects both notions of sampling inspection and AQL for the following main reasons:

1) The usual defect rate of 1% is considered too high and an important waste. A 1% defect rate means, for example, that 1000 out of 100,000 sold units are defective. This is unacceptable.

2) As JIT aims at one-piece-at-a-time production flow, 1% defect rate is difficult to apply or can mean 100% defective parts.[62]

3) Trained and professional inspectors can neither eliminate nor erase the defective occurrence. They cannot help attain a 100% good quality product. The quality problem cannot be solved by inspection. In fact, by the time the inspection is being carried out, the product is already good or defective.[63] JIT insists on the following point: you cannot inspect quality into the product; you have to build it in.[64] As JIT aims at zero detectives, i.e., a 100% quality products, JIT emphasizes

---

59 K. of Nippon Electric Company, cited by Sh. Shingo, *Study of Toyota Production System,* 1981, p.21. The company identified as Nippon Electric Company is now a JIT company

60 Buffa, E.S. and Rakesh K. Sarin, *Modern production operation management,* N.Y.: John Wiley & Sons, 19S7, p. 393–433

61 Reddy, J., and A. Berger, "Three essentials of product quality", *Harvard Business Review,* July/August 1983, p. 156

62 Haves R. H., "Why Japanese factories work", 1981, p.62

63 H. Ford (1988) believed strongly in inspection. Speaking of his manufacturing method, he wrote: "The key of our production is inspection. More than 3 percent of our entire force are inspectors…Every part in every stage of its production is inspected." p. 103

64 As regards the built-in quality, please refer to Gitlow, H. S. and P. T. Hertz, "Product defects and productivity", *Harvard Business Review,* Setout. 1983, p. 140; Hayes R.

not only a 100% inspection (total quality control at all levels of the production) but also the fact that the work should be done correctly in the first place.

However there is no doubt that for the worker, applying a 100% inspection must be too demanding and may force him to a situation of stress. JIT is aware of this and preaches the respect for the human being. That is why it imagined and invented some techniques to simplify the inspection.

In order to relieve the worker of the stress and continual efforts of thinking required by a 100% self-inspection, Toyota, the initiator of the JIT system, created a device known as foolproof or poka yoke.[65]

Concerning this poke yoke, Schonberger re-states its role:

> *As applied to quality control, bake yoke may refer to devices attached to machines to automatically check for abnormals or for malfunctions (...) Some baka yoke automatically stop the machine when a problem is detected.*[66]

As one should have realized it, the emphasis is placed not on finding the defective but on preventing it.

At a JIT company like Toyota an autonomous machine or machine with a human intelligence is the one equipped with a poke yoke type device that gives it the faculty of 1) detecting a defective or any other abnormality, 2) bringing it to the light and 3) deciding whether to stop itself or the entire line or not.

One may wonder whether it is not too costly to equip machines with such devices known as poke yoke, fixed-position stopping devices or full work system. Under JIT, it should not be so since these devices should be home-made.

The immediate goal of autonomation is double. 1) It helps to eliminate the overproduction due to the difference of the production capacity of different machines on the same processing line. 2) It tries to eliminate the waste of (making) detectives.

H., "Why Japanese factories work", 1981, p.62; S. Konz, "Quality circles: Japanese success story", *Industrial Engineering*, October 1979, *p.24;* Sh. Shingo, *Study of Toyota Production System*, 1981, *p.22, 183*

65 The following designations mean the same and thus interchangeable: Poka Yoke(Shingo, *Toyota Production System*, 1981), Baka yoke (T. Ohno, *Toyota Production System* 1988), foolproof (Y. Monden, *Toyota Production System*, 1983) or mistake-proof devices (Hall, Zero inventories, 1983).

66 R. J. Schonberger, Production workers bear major quality responsibility in Japanese industry", *Industrial Engineering*, December 1982, p. 39

Autonomation presents some advantages. It assures the good quality of all the items manufactured and at a very low cost: no scrap, no re rework, no routine inspection and no warranty losses and no extra-investment since poke yoke devices are the results of the workers' ideas and experience (See Reddy & Berger, 1983. p. 153–159).[67] Another advantage of Autonomation is the fact that it makes machines independence from men. Machines work autonomously and call the man's attention when a problem occurs. Therefore, they help reduce the number of workers on the processing line. In a word, the autonomation objective is to assure the quality of each product and that of the lot. It contributes to the reduction of the cost by prevention of overproduction (work system), by producing 100% defect free products (self inspection), by reducing the man-hour (reduction of the number of the worker). Besides this, the autonomation relieves the worker of the stress of always paying attention not to make a mistake and not to allow the production of a defective item. Such a worker can concentrate his entire energy on the work operations only and on the way to improve them.

3.6. Total Preventive Maintenance (TPM)

Maintenance at a JIT company like Toyota, though machines are so equipped that they stop automatically and that workers are also urged to stop the line in case of abnormality, the ultimate wish of the supervisors is that machines should not be stopped. So when a machine stops, the cause should be investigated thoroughly so that the recurrence must be prevented.

The prevention of the defective production is more emphasized than breakdown fixing.[68] That is why, besides actions taken to prevent defectives, machines undergo a preventive maintenance to avoid breakdowns. It is necessary to understand that, for JIT, 'defective' means more than defective items. It includes also defective work or defective functioning. Consequently, it is erroneous to emphasize only the quality of the product. That quality may be obtained by reworking the defective parts but at an extra cost:

Defects are not free: somebody makes defects and gets paid for making them.

---

67   See J Reddy and A. Berger, "Three essentials of product quality", *Harvard Business Review*, July/August 1983, pp. 153–159

68   See Walleigh, R. and M. Sepehri, "H-P Division programs reduce cycle times, set stage for ongoing process improvements", *Industrial Engineering*, March 1986. pp. 7

If a substantial proportion of the workforce corrects defects, the company is paying to correct defects as well as to make them.[69]

Fegenbaum points out that in some conventional manufacturing systems where rework on defective happens continuously there exist two plants in one plant: he calls the plant for rework a hidden plant that in some cases consumes from 15% to 40% of the production capacity.[70]

How to implement TPM program? Simmer et al. speaking of an actual case of TPM state clearly its goal, the moment it is carried out and the people in its charge:

A total preventive maintenance program was also established to minimize equipment breakdowns ... The program was set up to have the maintenance organization check every machine at specified intervals. This was done during slow period so that the work could be arranged to accommodate the preventive check. Operators were trained to perform minor checks every day. There was also a time limit established for maintenance as to determine the cause of a machine breakdown.[71]

A typical JIT company has two shifts, the third one being reserved for preventive maintenance.[72] Such a commitment to the preventive maintenance keeps machines up and functioning properly, which prevents defects and improves the process yields considerably. Schonberger compares this kind of maintenance to the air-plane-pilot style machine checking.[73]

---

69 Gitlow, H. S. and P. T. Hertz, "Product defects and productivity', *Harvard Business Review,* Sept./Oct. 1983, p 132

70 Fegenbaum quoted by R.J. Schonberger "Production workers bear major quality responsibility in Japanese industry", *Industrial Engineering,* December 1982, p.36

71 Simers, D., John Priest and Jack Gary, "Just-in-time techniques in process manufacturing reduced lead time, cost; raise productivity, quality", *Industrial Engineering,* January 1980, p. 22

72 That is the case of Toyota which I visited many times

73 R.J. Schonberger,"Production workers bear major quality responsibility in Japanese industry", *Industrial Engineering,* December, p.39

3.7. SS/QCC as frameworks of improvement activities

The search for the better quality, the reduction of costs and the elimination of waste are the main objectives of the improvement activities. JIT companies are committed to the philosophy of continuous improvement.

The total commitment to continuous Improvement may be analyzed in the following elements that Nemoto calls improvement principles:[74]

3.7.1. Improvement principles

(1) Seeds for improvement are limitless

Failures and mistakes are seeds for improvement. Every time that a failure occurs on the part of the machine, the process or the worker's operations, that failure becomes a seed for improvement.

Every failure or mistake must trigger thus an improvement plan or activity in order to prevent its recurrence. Therefore failures/mistakes should not be hidden; they must be brought to the attention of everyone working in the factory.

JIT is a dynamic system. It is never satisfied of itself and is in a continual need of improvements in order to reach its unreachable goals some of which are: zero setups or zero setup times; zero detectives i.e. no defective work and no defective items; zero inventories; one-piece-at-a-time flow or one flow processing line; the reduction of the work force to its strict minimum which may be just one operator: the minimum number of kanban; zero production lead times etc.

(2) Always refine yourself your machines, the new as well as the old ones

There are two kinds of machines that a company can be using: old machines and new machines.

Old machines should not be discarded on the ground of accounting principles as to which they may be considered depreciated or amortized. According to Ohno, once equipment is amortized, the rest is profit and he argues that with a better idea, a slight modification will make it possible to use the old equipment efficiently since the old equipment will contain new parts or new ideas of the latest technology.[75] The righteousness of Ohno's idea is confirmed by the following observation by Lesnet about a company that switched to JIT and improved its machines:

---

74   See Nemoto, M., *Total quality control for management: strategies and techniques from Toyota and Tovoda Gosei,* Englewood Cliffs, NJ: Prentice Hall, 1987

75   T. Ohno, *Workplace management* (translated by A.P. Dillon), Cambridge, Mass.: Productivity Press, 1988, p. 63–64

The JIT approach involved no new technologies, just refinements of already proven idea.[76]

The most eloquent proof is given by Toyota No. 1 Kamigo plant which is in McLeod's terms[77] "probably the most efficient engine plant in the world" and of which Schonberger states the particularities:

The No 9 Kamigo plant is equipped with twenty-year-old machines from Amerlca: machines tools from Cincinnati Milcron and both Excel_O and Cross transfer lanes. Over the years, the machines have been re-modified so they don't miss a beat. Limit switches and electric eyes check, count and index. If a machine makes a bad part or breaks down the giant overhead electric Jidoka signal lights up and summons help to fix the problem right away.[78]

Table 3–3 Comparison of Engine Plants

	Toyota Kamigo No. 9	Chrysler Trenton	Ford Dearborn
Products	2.4 L4-cyl.	2.2 4-cyl.	1.6 L4-cyl.
	2.4 L4-cyl.	inclus. Turbo	HO; turbo; EFI
Plant size (sq. ft.)	310 000	2 200 000	2 200 000
hourly employment	180	2 250	1 360
Line rate (per day)	1 500	3 200	1 960
Labor hours/engine	0.96	5.6	5.5
Shifts	2	2	1 assembly
			2 machining
Inventory (average)	4–5 hrs	2.5–5 days	9.3 days
Robots	None	5	NA

Source: R. J. Schonberger, 1986, p. 59

The efficiency of improvement on old machines is shown in Table 3–3..

---

76   D. E. Lesnet, "Facility found that means of implementing quick die changes were readily at hand", *Industrial Engineering*. November 1983, p.50

77   McElroy quoted by R. J. Schonberger, *World class manufacturing: the lessons of simplicity applied*, 1986, p.58

78   R J. Schonberger, *World class manufacturing: the lessons of simplicity applied*, N. Y.: The Free Press, 19S6, p 59

The greatest lesson for a manufacturer to bear in mind is that he must keep refining the old equipment instead of discarding them hastily as amortized.

It is worth noting that for JIT, not only old machines but new machines also, once purchased, need improvement.[79] JIT requires that before introducing new machines, you must make sure that people who will use them will be able to modify/improve them in order to suit your specific requirements. Otherwise, the workers with no capacity for improvement will simply end up being, as says Ohno,[80] slaves to the machine. In such a situation. the work force gets in Hayes' terms deskilled:

> *Our whole philosophy has been to distill our work force through automation. So we end up having relatively unskilled people overseeing highly sophisticated machines The Japanese put highly skilled people together with highly sophisticated machines and end up with something better than us.*[81]

The most important lesson to learn here is that improvement does not consist so much in bringing in the latest machines with the latest technologies but essentially in the ability of the workers to modify and adapt machines, the new as well as the old ones so that they can fit the special requirements of the company. In other words, the work force has to understand the process fully and so well that they must be able to improve it, to document it and to take its responsibilities.[82] They must not at any rate become slaves of machines: they ought to be their masters.

(3) The workers' expertise is the key to improvement

As it has been stated just above, improvement is not something brought about overnight. It is not something brought from outside.

Only the worker's knowledge is the key to improvement since they are the very experts. Hayes confirms this point of view in his marvelous article, "Why Japanese factories work".

---

79   Shinohara reports the case of Kibun, a Japanese company that switched to JIT and modified new machines to fit its needs, see Shinohara, I., *NPS (New Production System): JIT crossing industry boundaries*, Cambridge, Mass., Productivity Press, 1988, p. 137

80   T. Ohno, The Workplace management, 1988, p. 123–124

81   Hayes R. H., "Why Japanese factories work", p.64

82   Walleigh, R. and M. Sepehri, "H-P Division programs reduce cycle times, set stage for ongoing process improvements", 1986, p. 77

Schonberger describes vividly the way workers develop their expertise and the way that expertise can be tapped:

> *(…) many of the best ideas arise out of the everyday observations of employees not the abstract analysts of student engineers. Semi-skilled operators have the hand-on experience to conceive of and install, say, simple machine to machine transfer chutes as rollers. They can rough out plans for warning devices and electrical switches that synchronize processes, even though engineers will generally be needed to refine the design. Tapping the minds of operators in these ways is like turning on improvement engine at each machine.*[83]

Therefore workers are the main source of successful ideas and methods for improving the manufacturing processes.[84]

(4) Management at all levels must be involved in improvement

Management must also participate actively in improvement activities and set thus the pace for everyone. Unless they do so, programs of improvement will lead nowhere.

The suggestion system (SS) and the quality control circles (QCC) can be viewed as the formal framework within which improvement activities are actually conducted.

3.7.2. Suggestions systems (SS)

The purpose of SS is to tap the worker's expertise and knowledge and to apply them in improving the quality of the product, the work, the process and machinery. All that results in reducing the manufacturing costs. Suggestions can be made or introduced either by an individual worker or by a small group known as QCC.

At Toyota, suggestions or proposals for improvement are recorded at the plant head office and every month on a scheduled date they are examined.

Adopted suggestions for improvement are immediately announced in the Toyota Newspaper. Authors of successful suggestions are commended.

---

83 R. J. Schonberger, "Frugal manufacturing", *Harvard Business Review*, Sept.'Oct. 1987, p. 100

84 Sepehri, M., "Competition requires management to focus attention on manufacturing", *Industrial Engineering*, May 1987, pp. 6–8

3.7.3. Quality Control Circle (QCC)

Hall[85] defines the QCC as a group of 5–12 persons who work together, meeting voluntarily to improve the quality or to resolve other problems. Hall's definition is correct though it does not strictly fit many of the Japanese QCC. At some Japanese JIT companies like Toyota workers are required to take part in the QCC meeting; it is not a voluntary activity; it is a duty. So the QCC structure is strongly connected to the formal organization. That is why everybody must participate in its activities. Its goals are not confined to the quality of the product only. QCC aims at cost reduction, maintenance problems, the safety of the work for the worker, industrial pollution, etc. QCC members meet twice or three times a month for 30 to 60 minutes.

To sum up, there are many types of improvement: operational improvement, equipment improvement, product improvement, process improvement, work force reduction as improvement, improvement in increasing the safety of the worker and the work place; etc.

---

85 Hall, R.W., Zero inventories, Homewood, Ill.: Dow Jones-Irwin, 1983, p. 20 & p. 173

Chapter 4 Survey-based study of JIT techniques in the small and mid-size enterprises

To what extent does the Japanese company use "Japanese" management and production techniques? A survey targeting Osaka enterprises was conducted in September/October 1991 and October 1992 thru February. Ten years later, i.e., in 2003, the same survey was realized but at a small scale in Wakayama Prefecture, which like Osaka, belongs to the Region of Kansai. The survey consisted in an investigation about the distribution of Japanese special management features and production methods in the small and mid-size manufacturing.

The second chapter which is also based on the survey was devoted only to Japanese management features. Its conclusion has led to the necessity of rethinking the concept of Japanese management because that concept, in its present academic acceptation, does not seem to cover the majority of the population of the Japanese manufacturing world.[1]

This chapter draws all its substance on the same survey but it focuses only on the production methods: To what extent does the Japanese small and mid-size manufacturing enterprise use the production methods which are said to make the success of the Japanese corporation in general? Those production techniques, let me recall it, are known as making the just-in-time (JIT) system. JIT is also referred to as "kanban hoshiki", non-stock or zero inventories production methods. For the sake of simplicity, only the term JIT will be used from now on. Following are the survey results on JIT.

4.1. Presentation of the survey findings

4.1.1. Does your company use JIT?

Does your company use JIT? 114 out of a total of 129 companies covered by the survey replied positively or negatively to the question. That represents 93.8% of the companies. Of that large majority, a very small minority consisting of only seven companies or 5.79% do use JIT.

---

1   See L. Kupanhy, "Japanese management in the small and mid-size manufacturing- a survey", *Keiei Kenkyu* (The Business Review), Vol.43, No.3, September, 1992

Among the seven one company had people in charge of JIT implementation undergo a special training. Unfortunately the company did not specify as it was asked whether the training took place in or outside the company and who assured it. The other six did not have people in charge of the JIT implementation undergo a special training.

Concerning the period of time, only one company said it has been using JIT production method for 17 years and that it took a year to implement the system.

Anyway, a close look into the seven companies which use JIT suggests at least two things. First, it seems not necessary to undergo a special training in order to try to switch to the JIT production methods. Second, it may take a year or so in order to get the system put in place.[2] It is worth noting that a training program would have the advantage of speeding up the JIT introduction and help avoid going ahead by trials and errors.

The main lesson imposed by the great number of "No" is the fact that JIT seems not to be in use in the small and mid-size manufacturing enterprises. Consequently, a lot of sub-questions about JIT became regrettably obsolete. Among them the following ones are worth mentioning: Does JIT have a special name at your company (at Toyota, it is called TPS, at Kawasaki motors, KPS,[3] at Yamaha Motor, PYMAC (Pan Yamaha Manufacturing Control) though it is referred to as Synchro-MRP by R.W. Hall.[4] What was is the product defective rate before and after JIT introduction?[5]

Based on the negative findings from that starting question, one might hastily conclude that JIT is not practiced at all in the Small manufacturing and that there is no need to continue paying further attention to other JIT elements investigated. I think that would be an error. That is one of the points this chapter will try to show. In fact, one question kept constantly arising in my mind after considering those results about JIT itself: Does it mean that JIT used by big corporations is 100% supported by traditional production methods? To understand the

---

2   Sh. Shingo, *Study of Toyota Production System from industrial engineering point of view.* 1981, pp. 329–332

3   When I visited a Kawasaki factory, in Kobe that makes trains, I was told that that plant does not use JIT but that the division of Kawasaki Company that makes motor-bikes uses its own JIT system called Kawasaki Production System or KPS

4   R.W.Hall who wrote about it and called it Synchro MRP (See Hall, R.W., "Synchro MRP: combining kanban and MRP: the Yamaha PYMAC system", in Y. Monden, ed., *Applying just in time: The American/Japanese experience*, 1936, pp. 18–31)

5   To those questions one should add the following that were partly dealt with in the seven JIT companies: Do those in charge of introducing JIT undergo in your company a special training. If yes, within the company or outside the company?

pertinence of that question, it would be necessary to recall the following facts concerning the manufacturing world of the Japanese industry in general. First, most Japanese small and mid-size manufacturing enterprises are subcontractors.[6] Second, those subcontractors are the basis of the success of the big corporations since they manufacture more than 70% of the parts used in the products made by big corporations.[7] Third, in some cases they are the actual manufacturers while their parent companies look really like, in Sakai's terms, assembly lines and traders.[8] Fourth, JIT also helps eliminate most of the manufacturing wastes so that some big corporations help their subcontractors switch to or implement JIT methods.[9]

That is why I thought concluding at that stage of the survey that JIT does not exist at all in the small and mid-size manufacturing would be too hasty. It seems necessary to examine the other questions relating to JIT elements. Doing so might shed some light on the real status of JIT in the small and mid-size manufacturing.

Putting aside technical details of the JIT system, the latter is known to feature among many others the following characteristics: 1) quality control circles (QCC), 2) suggestions system(SS), 3) continual improvement, 4) mixed model production (MP) i.e., making of different kinds of products on the same production line, 5) short lead time, 6) total preventive (or productive) maintenance' (TPM), 7) multi-machine manning workers (MMW) or multi-process handling worker, 8) Non-stock (NS) or (near) zero stocks (ZS), 9) on-the-job training (OJT), 10) job rotation (JR).

Among those elements, one should have realized that there are Japanese management features, i.e., OJT and job rotations. Those two management-related elements were already dealt with in the second chapter and will not be paid any more attention to here.

I will present the survey findings about JIT first and about the supplier relationship next.

---

6   See Chusho Kigyo Cho (ed.), *Zu de miru chusho kigyo hakusho*, Tokyo, Okura-sho, *1992;* Ogawa, U., "Nihon no shita-uke soshiki. Sono genjitsu to tenho", Soshiki Kagaku vol. *24,* No. 3, 1991, pp. 28–39

7   I have visited a number of big Japanese manufacturing companies and none of them makes more than 30% of the parts needed for the products they make.

8   See K. Sakai, "The feudal world of Japanese manufacturing", *Harvard Business Review*, November/December, 1990, pp. 33–51

9   According to Monden and Nemoto, Toyota help its supplier implement JIT

4.1.2. JIT elements

4.1.2.1. Quality control circles (QCC)

Does your company use QCC? 118 of a total of 129 companies covered by the survey replied positively or negatively to the question. That represents 91.47% of the companies. That means that the result may be well extrapolated to represent the whole sample first and, at a higher level, the area covered by the survey. Among that large majority (91.47%), a large minority that represents 48.31% of the companies said they do use QCC (see Figure 4–1).

Figure 4–1 Distribution of QCC companies depending on who takes part in QCC activities

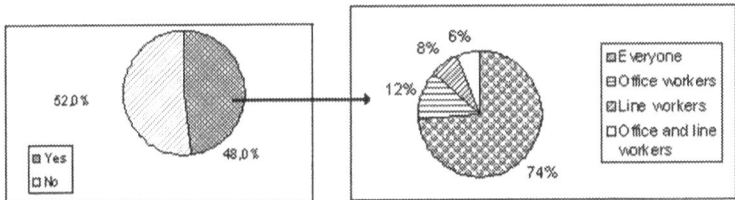

I am interested in that large minority within which the following questions will be examined: 1) Who takes part in QCC? 2) Are QCC activities conducted after work? 3) Can QCC activities be considered overtime 4) How long does a QCC meeting last?

1. Who takes part in the QCC activities?

Of the 57 companies which have QCC, 50 companies or 87.72% did give details as to who is involved in such activities. In 37 companies, the QCC involves everyone; that is what is referred to as total quality control (TQC) or company wide quality control (CWQC).[10] Six companies mentioned office workers as those concerned by QCC while four companies said those activities are primarily reserved for line workers. For three companies, QCC is a matter of office and line workers. Details are given in Figure 4–1.

---

10 According to M. Imai, "TQC is often understood in the West as part of QC activities, and it has often been thought to be the job for quality control engineers. Given the danger that the name TQC might be misleading and might fail to clearly communicate the scope of Japanese-style TQC, the term company-wide quality control (CWQC) was coined as a more precise term to use in explaining Japanese quality control to overseas observers", (Kaizen. The key to Japan's competitive success, 1991, p.43),

2. Are QCC activities conducted after work?

Among the 57 companies which have QCC, 54 answered the question. When do QCC activities take place? Are they conducted after work? Concerning the first question, 50 companies out of 54, i.e., 92.59% answered. Among them, the average number of companies which have such activities during the working hours is 50%.

This reminds me of Daikin, a major air-conditioning manufacturing company where production operators stop working for a half hour and devote that time to QCC activities.[11]

3. QCC activities as overtime

If QCC activities take place after work, is the time devoted to them considered overtired and duly paid)?

As to the fact that QCC may be considered as overtime, 51 companies out of 54 or 94.44% answered the question. Of those conducting QCC activities after work, 21 or (41.18%) companies consider such activities to be over-time while the remaining 58.82%, i.e., 30 out of 51 companies do not.

4. QCC activities' frequency and duration.

As to the frequency of the QCC activities, almost they are held on a weekly basis. 27 companies out the 54 which said to hold QCC activities did indicate the time a QCC meeting lasts. The QCC meeting time ranges from 10 to 150 minutes. As many companies indicated both the minimal and maximal time, the minimal average time is 54.259 minutes while the maximal time, the average is 60.926 minutes.

The mode for both the minimum and maximum is 60 minutes. Let me recall the statistical definition of the mode. It is "the value in a frequency distribution which occurs most frequently.[12] The total cumulative time devoted to QCC activities ranges from 1465 minutes 1656 minutes a week.

The percentage of companies whose minimal activities are equal to or less than 60 minutes is 88.889%, i.e., 24 companies while 22 companies or 79.481% devote a maximal time not exceeding 60 minutes.

---

11   I visited two factories of that company in Sakai City, Osaka Prefecture (in 1990)

12   The Living Webster. Encyclopedic Dictionary of the English Language, Charles E, Tuttle Co., Tokyo, 1975

4.1.2.2. Suggestions Systems (SS)

1. Do you use the suggestion system (SS)?

Figure 4–2 Number of companies featuring SS

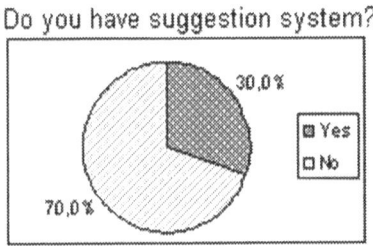

Does your company use SS? 115 of the total of 129 companies covered by the survey replied positively or negatively' to the question. That represents 89.15% of the companies as shown in Figure 4–2

That means that the result may well be extrapolated to represent the whole sample first and at a higher level to cover the area covered by the survey. Among that large majority (89.15%), a majority representing 69.57% of the companies which answered that question said they do use SS. The following lines will focus on those 69.57 %, i.e., 80 companies out of 115.

2. Number of suggestion per company per year

Among the 80 companies that have SS, 49 companies or 61.25% indicated the number of suggestions they are getting per year.

The minimal number of suggestions ranges from one to five thousand and the maximal number from two to five thousand. In a word, the overall range of the number of suggestions goes from 1 to 5000. The total number of suggestions varies from 11612 to 11698.

The minimal average number of suggestions is 236.98 and the average for the maximum number of suggestion stands at 238.735.

For both the maximum and minimum number of suggestions, the mode is situated at 20 suggestions with a frequency of 8 and 6 for the minimum and the maximum number of suggestions respectively. Table 4–1 summarizes of the results.

Table 4–1 Summary of the number of suggestions

	Suggestions		Mean	Cumulative total	Mode (Freq.)	Number of companies
	From	To				
Minimum	1	5000	236.98	11612	20(8)	49
Maximum	2	5000	238.73	11698	20(8)	49

A close look at the cumulative frequency percentages and the number of suggestions (column 1 and column 4) shows that the number of 900, 1500, 1700, and particularly 5000 suggestions are too far apart from all the others and seem quite out of the normal range of representation. 45 companies have the number of suggestions per year within 350 (including the latter).

That means that excluding the four numbers would give a mean that reflects better the majority of companies. Besides, the sum of suggestions for the four values is 9100 while that of the remaining 45 companies is between 2512 and 2598. Furthermore, the sum of the remaining 45 companies is even about half the value of 5000 suggestions. Not taking those extreme cases into consideration would yield the following means for the minimum and the maximum number of suggestions: 2512/45 = 55 and 2598/45 = 57 I think that the mean that ranges from 55 to 57 suggestions per company per year is more representative.

3. Number of suggestion per individual per year

53 companies among the 80 that have suggestion systems gave some figures as to the average number of suggestions they collect per individual per year. That represents 66.25%.

The number of individual suggestions ranges from zero to 85. The mode for the minimum individual number of suggestions is one with a frequency recurrence of eleven. The value of two has the second highest frequency of 10. Concerning the maximum number of suggestions, the mode is two with a frequency of 11. In the second position, there are two values: one and three that have each a frequency of 8. Three companies get each on average 30, 70, 85 individual suggestions respectively. That seems to be so out of the normal range that they may be excluded in order to extrapolate the result. The bulk of companies have the number suggestion per individual varying between zero and 10

4. Yen value of suggestions

Those suggestions are worth some savings, of course. Of the 80 companies which have SS, 26 companies or 32.5% stated the amount of yen suggestions are worth.

The yen value of suggestions ranges from 1.000 yen (in 3 companies) to 5 million yen (one company). The average varies from 696,538 to 718,077 yen. Altogether, the total values for those suggestions ranges from 18,110,000 to 18.670.000 yen as one can be seen in Table 4–2:

Table 4–2 Summary of yen value of suggestions

	Suggestions Yen value		Mean	Cumul. total	Mode (Freq.)	Number of companies
	From	To				
Minimum	1000	10 M	696538	18.11M	100000(6)	26
Maximum	1000	10 M	718007	18.67M	100000(6)	26

Values of one million yen, five million yen and ten million yen of which the total frequency is four companies or 15.39% are so far from others that they can be omitted in order to find a more representative mean. In fact, out of a total ranging from 18.110.000 to 18,670.000 yen, suggestions for the four companies are worth 17.000.000 yens while the suggestions values of the remaining 22 companies give a sum ranging from 1,110,000 to 1,670,000. The average number of suggestions for the 24 remaining companies is to be situated between 504,545 and 75.909 yen.

4.1.2.3. Improvement

Figure 4–3 Number of companies with improvement

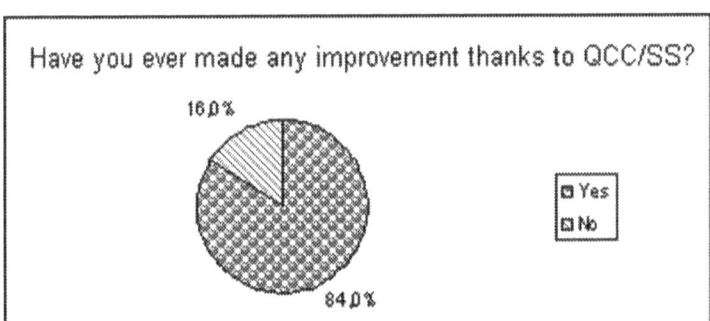

The questionnaire wanted to know whether some improvements were realized thanks to QCC and/or SS activities. 86 companies dealt with the question about improvement and the remaining 43 paid no attention to it. The number of respondents represents 66.67 % among which 83.7% or 72 companies made some kind of improvement. Fourteen companies or 16.21% said no improvements were accomplished (Figure 4–3)

I will dig into the group of those 72 companies who did make some kind of improvement in order to know their authors. 93.06% of the 72 companies, i.e., 67 companies answered the question concerning the architects of improvement.

The participation or contribution to improvement has the following configuration. Out of the total of 67 companies, engineers participated in 38 companies, operators contributed in 47 companies and consultants in 5 companies as shown in Figure 4–4. That represents receptively 56.71%, 70.15% and 7.46%.

Figure 4–4 Distribution of companies according to improvement contributors or authors

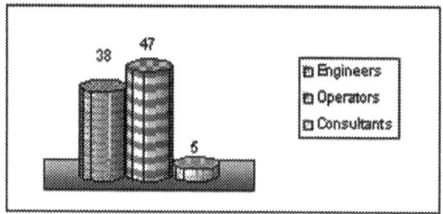

As one would have remarked, the total frequency (38+47+38) of 90 exceed the total number of companies which stand at 67. The same is also true for the total of their percentages (56.71 + 70.15 + 7.46 = 134.32) that exceeds 100%. This is due to the fact that some companies are counted twice or three times if more than one category of workers has been involved in improvement. When frequencies of replies are re-considered with regard to the total number of frequency values of 90, the respective frequencies will represent 42.22, 52.22 and 5.56%.

A close look at the improvement by each category of employees shows that the most significant scores of improvement are dominated by engineers and/or line operators. The contribution from operators counts for more than half of the total number of cases.

Roughly speaking, of the companies which said to have had some improvement, more than half (52.22%) had those improvements realized by line workers only; the number of companies where engineers were architects of improvement represent only 42.22% (see Figure 4–4).

Figure 4–5 Distribution of companies according to detailed analysis of contributor types

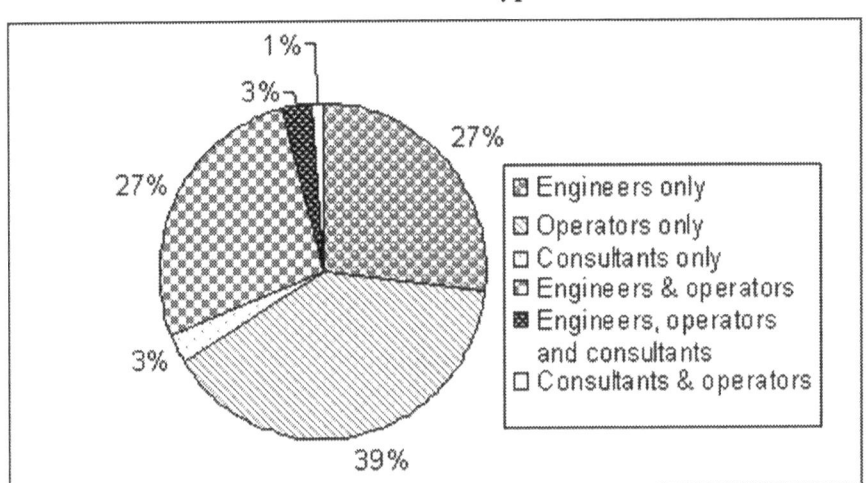

A strict observation gives the following results. Improvements by engineers only were made in 18 companies, those by line operators only in 26 companies, those by only consultants in 2 companies. That represents 26.87%, 38.81% and 2.99%. 18 other companies mentioned both engineers and line operators. In two companies only were improvement made by the three types of employees, i.e., engineers, line operators and consultants (Figure 4–5).

4.1.2.4. Mix production (MP)

75.19% of the respondents answered the question about the mix production, i.e., the production of different items on the same assembly line and the lead time. Of those, 94.85% produce many different types of products on the same production line (mass production of mix products) while only 5.15% produce a single product.

4.1.2.5. TPM

The total productive maintenance is supported in big corporations by a shift that is known as the maintenance shift. Usually that is the night shift.

Most companies covered by the survey did not seem interested in indicating the number of shifts. Out of a total of 129, eighty-nine (89) companies representing 69% did not answer the question. The remaining 40 companies represent a small minority of 31% in which a very large majority of 32 companies (out of 40) or 80% have only one shift. Six companies (15%) have two shifts among which one company has a shift devoted to TPM. Two companies out of 40 or 5% of the respondents have three shifts and one of the two has established a TPM shift. In other words, only two companies have a maintenance shift.

That suggests that TPM is not something one might expect to find widely spread in the small and mid-size manufacturing.

Do you use TPM? 73 companies or 56.59% replied to that question by yes or no.

When replies are grouped into two groups, one for yes, and the other for no, one realizes that TPM is almost not practiced: only 6 companies out of the 73 or 8.22% of the respondents do have TPM programs

Companies that gave no replies (56 companies) and those which said no (67 companies) were given the possibility to specify whether the TPM notion was unknown to them. Out of a total of 123 companies (129 minus 6 companies with TPM programs), 84 companies or 68.29% did not reply while the remaining 39 companies 31.7 1% did positively or negatively (see Figure 4–17). Among the latter, 4 companies or 10.26% know of TPM while for the other 35 or 89.74%, TPM was something unknown.

4.1.2.6. MMW

Concerning this point, companies were not asked whether they have multi-machine manning workers[13] or not. They had rather to indicate the minimum and the maximum number of machines or processes a worker on the production line can handle. And 41 companies or 31.78% specified the range of the number of machines or processes anyone of their line operators has to handle. I am going to focus on that small minority.

Table 4–3 Distribution of companies according to the number of machines or processes handled by operator

	Suggestions		Mean	Mode (Freq.)	Number of companies
	From	To			
Minimum	1	5	1.94	1(16)	36
Maximum	2	10	3.72	2 & 3(11)	36

Five companies out of those 41, i.e., 12.20% continue using the traditional method with a worker manning only one machine. In the remaining large majority of that small minority a line operator handles at least one machine or process.

---

13 Multi-machine manning worker (Abegglen & Stalk, Jr., 1987; Shingo, 1981), multi-process handling operator (Shingo 1981), multi—function worker (Monden, 1983, 1998), are just to be considered synonyms.

In the 36 companies that have multi-machine manning operators, the range of the minimal number of machines or processes an operator supervises goes from one to five (Table 4–4). As to the maximum number of machines or processes a line worker handles, it is between two and ten. The mode is one machine for the minimum with a frequency rate of 44.44%. For the maximum number of machines a worker can handle, the values of two and three machines per worker have the highest rating of 11 each or 30.56% among companies.

Though the computer could compute the average which is 1.94 machines for the minimum number and 3.72 for the maximum (Table 4–4), it has no practical meaning because only a weighted average would be significant. The latter is difficult to find here since I do not know exactly how many workers did handle how many machines per company.

4.1.2.7. STOCKS

The questionnaire dealt with the problem of stocks in two parts: stocks of finished goods and work in process (WIP) inventory.

Stocks of finished goods

Do you have stocks of finished goods? 90.70% of the companies replied to this question among which 67.525% said they have stocks of finished goods. Therefore, only a small of minority of the respondents, i.e., 32.48% are running their companies on the basis of what is known as a strict zero stock production method.

Of the 79 companies running stocks, 51 companies or 64.56% indicated the number of days their stocks are worth. Tables 4–4 summarize the findings regarding this point.

Table 4–4 Summary of stock value in days

	Suggestions range		Mod (Freq.)	Number of companies
	From	To		
Minimum	2	90	30 (9)	51
Maximum	2	180	30 (9)	51

Those stocks are worth from two days (in two companies) to 90 days (is excluded just one company with a stock worth 300 days). The mode is 30 days with a frequency of 9 (companies). 33 companies out of 51 or 64.71% have a minimal stock equal or inferior to the mode of 30 days. As for the maximum, the number of companies whose number of days is equal or inferior to 30 days is 30 or 58.82%.

WIP inventories

The just-in-time conception considers as a waste the work-in-process (WIP) inventory. 86.82% of the surveyed companies dealt with the questions about WIP. Of those, only 23.2 15% do not have WIP inventories at all while the large majority representing 76.79% do.

The volume of WIP inventories can be assessed, like stocks, in the number of days they are worth. 48 companies out of 86 (or 55.81%) that run WIP inventories gave the number of days their wip inventories are worth (Table 4–5).

And both the minimum and the maximum number of days of WIP vary from one to 180 days. Their mode is 30 days with a frequency of 9 (companies) in each case.

Concerning the minimum number of days, 38 companies out of 48 have WIP worth or less than the mode, i.e., 79.16%. As to the minimum, 36 companies out of 48 or 75% have stocks of 30 days or less.

Table 4–5 Summary of WIP value in days

	Suggestion range		Mod (Freq.)	Number of companies
	From	To		
Minimum	1	180	30 (9)	48
Maximum	1	180	30 (9)	48

4.1.3. Supplier's relationship

Japanese companies can be grouped into the following three categories. First, there are companies that have suppliers but are not suppliers themselves. Most of them occupy the top rank in the pyramidal structure of the Japanese organization of the relationship between companies. Second, at the bottom, one will find those companies which are suppliers but do not have suppliers. At last, between those two extremities, you have those that are suppliers and at the same time have suppliers.

JIT concerns also the supplying and delivering relationship.

4.1.3.1. Do you have suppliers?

A score number of 110 companies representing 85.27% of the sample size answered that question. Among those 85.27% %, 76.36%, i.e., 84 out of 110 companies have suppliers.

Number of s appliers

55 companies out of 84 (or 65.48%) which said to have suppliers gave the number of their suppliers. The number of suppliers range from one to 200 companies. The number of suppliers range from one to 200 companies. The mode is the value of 20 suppliers (in seven 7 companies) followed by 6 companies which have 30 suppliers each. On the average, companies have 31.73 suppliers

Percentage of supplier parts into finished product

What is the percent of the supplied parts in your product? 74 companies out of 84 indicated the percentage of the supplied parts in the product that bears their trade mark. The percentage of supplied parts varies from 1% (in one company to 100% (3 companies). The mode is 3% with a frequency 12 companies.

In big corporations, supplied parts count for at least 70%. Of the 74 companies, only nine companies or 12.16% have the number of parts supper that are equal to or more than 70%.

Daily, weekly or monthly delivery types

Of the 84 companies that said to have suppliers, 69 companies or 82.14% did specify the type of delivery on which is based their relation with the suppliers.

The 69 companies may thus be grouped into three categories as regards the delivery type of the parts they order from suppliers: there are daily deliveries (27.54 %), weekly deliveries (46.38%) and monthly deliveries (26.09%) (See Figure 4–6). As one sees it, most companies work on a weekly delivery basis as regards their relationship with suppliers.

Figure 4–6 Delivery type

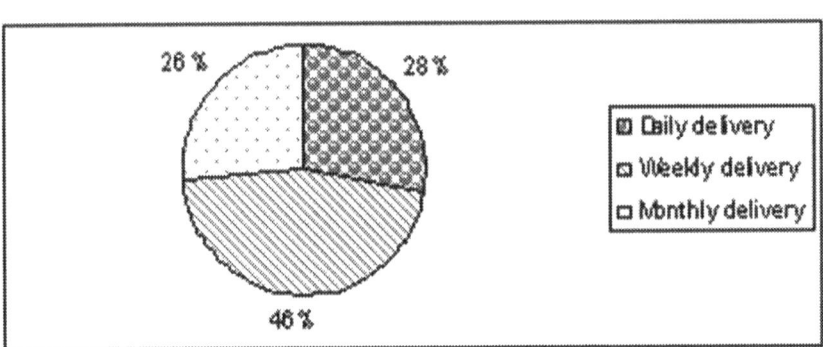

Part delivery frequencies

Do you get the parts you have ordered on a daily, weekly or monthly basis and how many times for each case? Figure 4–6 above has dealt with the first part of that question and the following lines will give the findings about the question's second part.

Of the 19 companies that operate on the basis of daily deliveries, the frequency of delivery ranges from once (that is the mode with a frequency of 11) to 15 times (two companies) a day.

As for those companies getting the ordered parts on a weekly basis, out of 32 companies, 11 companies receive the needed parts once a week and the other 11 companies twice a week. Seven companies get the purchased parts three times a week and the remaining 3 companies get the ordered parts 5 times a week. The range is from once to five times a week with the average situated at 2.16 times a week.

Eighteen (18) companies receive the parts they order on a monthly basis. The range goes from once to 1500 times a month. If one excludes the highest extreme value of 1500 times with the frequently of one, the highest frequency becomes five for the 17 remaining companies. Of those, six companies receive the parts once a month, eight companies twice a month (that is the mode) and the two others companies once and twice respectively. The mean is 1.83 times a month.

Delivery time

How long does it take from the time you have placed your order to the time the ordered parts will be delivered? 69 companies out of the 84 (or 82.14%) which said to have suppliers answered that question. Among those 69 companies, for some it takes just hours, for some others, it takes days, for a third group it takes weeks and for the last group, it takes months.

Figure 4–7 Distribution of companies according to the delivery time of ordered parts

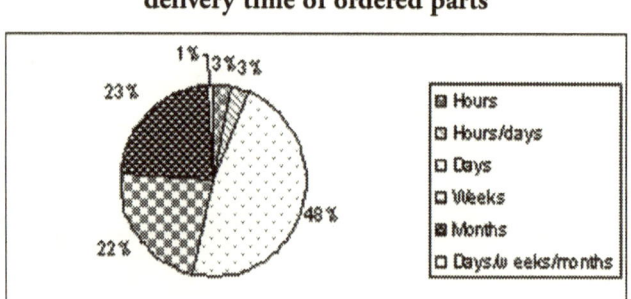

In fact, for 47.83% of the 69 companies, it takes days from the time they place an order to the time the order is served. In 21.74%, it takes weeks; and, months are needed in 23.19%. In 2.90%, it takes only hours and in 1.45%, the delivery time varies from hours to months (See Figure 4–7).

The hourly delivery system is practiced in two companies. In one of the two, it takes eight hours and in the other 9 hours from the time an order is placed to the moment that order will be served.

One company said it takes between three hours to 30 days for the delivery of the ordered parts. The company unfortunately did not indicate the number of its suppliers though it said suppliers' parts count for 10% of its product. It also said to get ordered products once a day.

For the companies in which it takes days before getting the ordered parts, the number of days ranges, as regards the minimum, from one to 60 days. There are two values with the highest frequency of seven, i.e., one day and 3 days. Among those companies, in 72.22% or 26 companies out of 36 (33 + 2 companies that have both hourly and daily delivery + one company that gets within days, weeks or months after the order is placed), the minimum time is equal to or less than 15 days. And 15 days can be considered as a short delivery time. The average for the minimum is 12.58%.

As for the maximum number of days, the delivery time varies from one day (4 companies or 11.11%) to 90 days (in 2 companies or 5.56%). The average is 18.69 days for the 26 companies whose delivery time is equal or inferior to 15 days. The value of three days, and 30 days have the highest frequency of five each. The number of companies with the delivery time equal to or less than 20 days represents 72.22%.

For 16 companies (15 + one company for which the delivered time takes days, weeks and months), weeks are needed to get the ordered parts from their suppliers. The length of time goes from one to 6 weeks (one company or 6.25%). One week is the mode for the minimum delivery time with a frequency of 6 while the value of two weeks is the mode for the maximum with a frequency of six too. The frequency of six represents 37.5% of the companies of that category.

The average varies from 2.19 to 2.25 weeks.

Of the 17 companies (16 + one company that gets ordered products within days, weeks or months) or 23.6% which get the ordered parts monthly after they have placed their orders, the number of months for the minimum varies from one month—which is also the mode with seven companies (out of 16 because one company did indicate the minimum) or 43.75% to 6 months (one company or 6.25%). The average is 2.13 months. The graph that makes up Figure 4–26 describes the delivery time that takes months.

Concerning the maximum number of months, the figures vary from one month (with 4 companies or 25%) to 6 months (3 Companies or 18.75%). The mode is 2 months with five companies or 31.25%. The average is 2.94 months. The total is 16 companies because one company did not indicate the maximum

4.1.3.2. Are you a supplier?

This question was answered by 92 companies or 71.32%. Of those a very small minority of 20.65% or 19 companies said they are suppliers.

The majority of that small minority, i.e., 12 but 19 companies or 63.16% indicated the number of the companies of which they are suppliers. Except two companies that supply two companies, all the others are suppliers to one and only to one company.

Figure 4–8 Distribution of companies according their response time

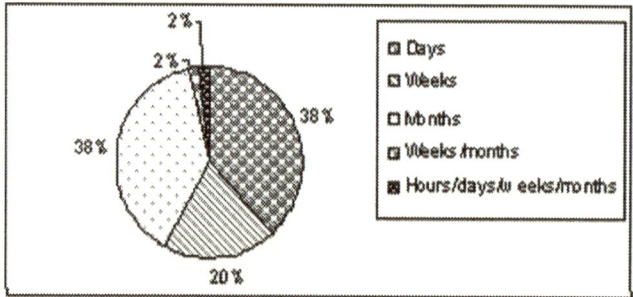

What is very interesting is the fact that as a rule, Japanese companies are sub-contractors only to one major company.

Concerning the response time to customer's orders, companies are grouped into the following categories: deliveries that take days, weeks or months.

66 compares out of 129 or 51.16% specified the type of the delivery. Of those companies, 25 companies or 37.88% deliver the product to the buyer within days, 19.70% or 13 companies within weeks and 39.40% within months. A company says it takes weeks and months and for another company, it has hourly, daily, weekly and/or monthly-based deliveries.

Daily based supplying time

Once an order has been received, the minimum number of days for supplying the parts varies from one to 50 days. The average is 13.58 days. One day value is the mode with a frequency of 7 out 26.

The maximum number of days varies from one (in 5 companies and that is also the mode) to 90 days (one company). The average stands at 20.92 days

Weekly-based supplying time

15 companies gave figures about the number of weeks it takes them to provide the buyer with the ordered products. The minimum number of weeks varies from one (with 7 companies and that is the mode) to 8 weeks.

The maximum number of weeks varies from one to 10 weeks (one company). The mode is two weeks and it has the frequency of 6.

The average ranges from 2 to 2.53 weeks.

Monthly-based supplying time

28 companies provided the needed information about the number of months it takes them to supply the parts for a received order. The minimum number of months varies from one (with 11 companies and that is the mode) to 8 months (one company) while the maximum number of months goes from one (that is also the mode with its highest frequency of 7) to 12 months (one company). The mean ranges from 2.64 to 3.89 months. If the values varying from 8 to 12 months mentioned by the same company are not taken into account, the mean would range from 2.5 to 3.59 months.

Concerning the minimum time, it is worth noting that the mode of one month with its frequency of 11 represents 39.28% and that 78.57% of the companies supply the parts at least within three months.

4.2. Discussion and conclusive suggestions: trends' detection

I do not have any intention to draw sharp conclusions based on the findings of the survey about JIT in the small and mid-size manufacturing enterprises. I am not even intending to extrapolate those findings. The attention will be not focused on broad generalizations of the findings for the Japanese companies such as: In Japan, only 5% of the small and mid-size manufacturing companies use JIT or most companies are running huge quantities of inventories. From the findings above, the reader may draw such kinds of conclusions if he/she deems them useful for some purpose. Besides, the data presented above will be used neither to criticize nor to raise doubt or questions about JIT in Japan as I did for the Japanese management. What is then the survey useful for? What is its objective? Was it not aiming at something? Did it not lead to anything new?

The first lesson worth mentioning that can be learnt from the survey findings is twofold. First, the assertion that the small and mid-size manufacturing cannot use JIT is questionable. Second, JIT features can be detected in the small and mid-size manufacturing. That leads to the suggestion that JIT can be considered as penetrating maybe very slowly but surely the small and mid-size manufacturing.

In fact, the approach to interpreting survey findings on JIT should differ from that about the management of the Japanese company. Japanese management is said to have a long tradition taking roots in the Japanese society and culture. That is why I thought it should be checked only if Japanese management features are or are not to be found in the small and mid-size manufacturing.

Concerning JIT, it can be said without fear of being contradicted that its history is very recent. Therefore JIT should be considered from the point of view of its possible expansion through the manufacturing world though most big corporations have already adopted it regardless of their production lines and the industry they belong to.

The experience of NPS group is a convincing proof that JIT may be applied to and is penetrating not only a type of industry but all kinds of the manufacturing industries regardless of the size of the corporations.[14]

When presenting data, some were discarded for being out of the normal range in order to find a mean that would be more representative. This is fine for a pure statistician or for the one who is interested in broad generalizations of the conclusions drawn from the survey findings.

Out-of-the-range results are of the highest interest for the present study in the sense that they indicate important orientations for the manufacturing world. They represent either the new directions of change or the strength of stagnation forces, i.e., forces of inertia, which by their very nature run counter the change of the status quo.

Forces oriented toward changes have as their model the performance of the best JIT companies at which they are aiming. The opposite forces or forces of inertia are represented by conventional manufacturing systems. A particular effort will be made to detect trends' indicators of those two types of forces. And where possible, some lessons may be drawn for a foreign company as to which options look better in order to implement JIT smoothly.

4.2.1. QCC, SS and improvement

Of the 129 companies covered by the survey, 57 have QCC and 61 do not. As one can see it, the forces of changes are almost equal to those of stagnation.

37 companies out of 50 said QCC activities involve everyone. That reminds of Toyota where they are part of everyone's normal duty. QCC activities are conducted after work (in 50% of the companies) or during the work time.

---

14 See I. Shinohara, 1) *NPS seisan shoshiki: fumetsu no keiei*, Tokyo: Toyo Keizai Shimbunsha, 1989; 2) *NPS no kiseki: Be-ru o nuida seisan hoshiki*, Tokyo: Toyo Keizai Shimbunsha, 1985. According to him, NPS is a consulting group that is initiating companies to JIT and there are over 400 hundred companies on the waiting list.

A foreign company trying to implement QCC would better think of conducting such activities during the working hours. If conducted out of working hours, such activities should be considered as overtime as does the minority of surveyed companies.

No best suggestions could be made about the length of time for a QCC meeting: depending on the good or advantage the company is getting from QCC activities, both extremities of 10 and 150 minutes sound acceptable.

If 10 minutes seem too short, the time may be extended and if 150 minutes look too long so that there is a lot of "waiting time" for ideas to come, the time may be shortened. As to the frequency of QCC meetings, once-a-week frequency may be a good practice, as is the case for most companies.

The number of suggestions per year ranges from one to 5000 though most companies (34 companies out of 49) are getting 50 suggestions or less. The minimal number of one suggestion per year (in 2 companies, see Table 4–1) should be considered as the starting point which should not last for a long time. That is almost the zero level. The numbers 5000, 1500, 1700 and 900 suggestions, each with a frequency of one, which were excluded in order to find a more representative mean (average) should be regarded as showing the trends or the new orientations of companies that are succeeding in implementing JIT. In fact, big corporations are collecting millions of suggestions (see Table 3–4).

Table 4–6 The ten most active kaizen programs in Japan, 1990

Company	Total Suggestions	Ideas per person
1. Kawasaki HI	6 980 870	426.5
2. Nissan	6 043 344	126.9
3. Toshiba	4 166 864	76.9
4. Matsushita	4 114 398	43.7
5. Mazda	2 417 264	113.0
6. Toyota	2 003 646	35.0[13]
7. Otsu tire	1 475 707	1185.3
8. Nihon Victor	1 247 523	83.1
9. Nissan Diesel	1 169 745	226.8
10. Fuji HI	998 359	88.1

Source: The Japan Human Relations Association, Summary of Japanese Suggestions Activities Survey, 1991. See S. Alan, G. Robinson and Dean M. Schroeder, "Training, continuous improvement and human relations: the US TWI programs and the Japanese Style", California Management Review, Vol. 35. No. 2 Winter 1993, p. 35–57[15]

The least performing companies get less than one suggestion per individual while the three most performing ones have respectively 30, 70 and 85 suggestions per individual. The company getting 85 suggestions is almost as good as some big corporations (see Table 4–6.).

Table 4–7 Yen value of suggestions at some companies

Period	Company	Investment in SS	Cost savings (yield)
1978–82	Nissan Chemicals	125,000	¥ 600,000
1981;1987	Nissan Chemicals	160,000	¥ 600,000
1983	Canon	250,000	¥ 19,300,000
1986	Toyota	NA	$ 230,000,000

Source: Adapted from Imai, 1991, pp. 107–120; Kenichi Omae, 1986

The company benefiting the most from suggestions earned 5 million yen. Big corporations with tens of thousands of workers and millions of suggestions earn hundreds of millions[16] (see Table 4–29). The amount of 5 million yen was excluded in order to satisfy the requirements of the pure statistician who sticks to the meaningless of a representative mean. But, in a field under change like the manufacturing sector facing the challenge of JIT, the figure is full of meaning since it tells anyone wanting to implement JIT that suggestions are a source of revenues. The clear message is that a company can earn millions from the ideas of its work force.

Improvements in the JIT production environment are made through QCC and/or suggestions systems. I am going to try to see whether in the field of small and mid-size manufacturing enterprises, there exists some kind of relationship between improvement, QCC and SS.

4.2.1.1. Improvement & QCC

It has been reported that in 72 companies improvements have been realized (Figure 4–12 A) and that 57 companies have QCC (Figure 4–2 A). Among the 57 com-

---

15   It is curious to see that there is a decline in suggestions at Toyota. In 1986, there were 2,650,000 suggestions in total averaging 48 suggestions per employee (*Production at Toyota.—Our basic philosophy*, Toyota TMC, without data p. 27

16   According to K. Ohmae, Toyota was getting suggestions worth *$230* million (*The mind of tile strategist*, N.Y.: Penguin Books, 1983, p. 207)

panies with QCC, 51 or 89.47% have made some improvements. In other words, 51 companies out of 72 (or 70.83%) which made improvements have QCC

4.2.1.2. Improvement & SS

Of the 80 companies that have established the SS (see Figure 4–8), 67 companies (or 83.75%) have succeeded in realizing some improvements. In other words, out of 72 companies accrediting themselves with improvement, 67 companies or 93.06% have SS.

4.2.1.3. Improvement, QCC & SS

49 companies have both QCC and SS. It means that 49 out of 57 companies (or 85.96%) with QCC have SS and 49 out of 80 companies or 61.25% with SS have QCC too.

Of the 49 companies which have both QCC and SS, 47 companies or 95.91% have experienced some kind of improvement.

It is worth noting the fact that all the 72 cases of companies with improvement were companies with QCC or SS or both QCC and SS. Not a single company without either of those organizations mentioned to have made an improvement.

4.2.1.4. Improvement architects

One would have surely realized that the base of improvement seems to be the line workers. In fact, the majority of companies with improvement are those where line workers contributed.

Though I know from my experience of the JIT study that line workers are the real architects of improvement, I can not here confirm the same. In my mind the main problem is the following: Does the quantitative superiority of the number of companies with improvement by line workers mean that in the whole those cases of improvement are also qualitatively superior? If I have to keep my trust in JIT theories and practice, the importance of line workers cannot be neglected and is out any doubt.[17] But for the case of the survey, one should not forget that the survey focused mainly on the number of companies that have made some kinds of improvement and not on the nature or value of improvement itself which is much close to the field of industrial engineering or value engineering.

---

17 See D. A. Garvin, "Quality on the line", *Harvard Business Review*, Sept./Oct. 1988, pp. 65–75; R. J. Schonberger, "Production workers bear major quality responsibility in Japanese industry", 1982, pp. 34–40

However, in order to get glimpse of the real architects of improvement, I have tried to find a correlation between improvements, their yen values, their authors and the number of companies in which they occurred. The yen values of improvement were obtained indirectly by the yen values of suggestions because an improvement is a result of applying suggestions for improvement.

24 companies out of the only 26 that indicated the yen values of their suggestions will be considered because two companies whose total yen values vary from 150,000 to 200.000 yen will not be taken into account. In fact, one of the two did not answer the question about improvement while the other said it made no improvement.

Table 4–8 Total value of improvement contribution by job category

Job category	Number of companies	Range of yen value contribution	
		From	To
Engineers	12	16,243,000	16,653,000
Operators	17	16,747,000	17,257,000
Consultants	1	30,000	30,000
Others	1	100,000	100,000
Total	31	33,120,000	34,040,000

As for the Yen values of improvement, engineers contributed in 12 companies for a total amount varying between 16,243,000 and 16.653,000 yen, operators in 17 companies for a sum situated between 16,747,000 and 17,257,000 yen. Consultants contributed to improvement in one company for a total of 30,000 yen. As one sees, the total number of contribution and that of companies in which those improvement values are mentioned exceed the total number of amount of yen contribution and that of companies. That can be seen in comparing Table 4–7 with Table 4–2.

Table 4–29 shows the detailed and exclusive contributions by each category of employees. In six companies, contribution to improvement was by engineers only and that represents an amount of 1,113,000 yen. In 10 companies, only operators did make improvements worth 1,587,000–1,687,000 yen. There is not a single case of improvement by consultants only. In the nine remaining companies, the contribution was by both operators & engineers or operators & consultants (see Table 4–8).

Table 4–9 Exclusive contribution by each category of workers

Job category	Number of companies	Range of yen value contribution	
		From	To
Engineers only	6	1,113,000	1,113,000
Operators only	10	1,587,000	1,687,000
Consultants only	0	0	0
Eng. & operat.	6	15,130,000	15,540,000
Operat. & consult.	1	30,000	30,000
Others	1	100,000	100,000

If one subtracts the three extreme values of 10,000,000; 5,000,000 and 1.000,000 yen the contribution by both operators and engineers falls to 130.000 yen and 540.000 yen respectively. But everybody knows that there is no need to do so because this indicates an orientation toward progress.

Consultants do not seem to play an important role in improvement contribution, both in terms of improvement value and number. In fact, it seems that the least involved people concerning improvements companies have realized are consultants. Is that due to the fact that because of the lack of sufficient funds small companies usually suffer from, the small and mid-size manufacturing can hardly afford to hire consultants? Or does it mean that consultants do not play any important role in improvements made at the work place? Both options are and remain possible because either is worth defending. .

Because there are not enough data, I would wise not to draw any firm conclusion about the small and mid-size manufacturing concerning the real architects of improvement between engineers, line workers and consultants though Table 4–29 suggests strongly the importance of those people who have their hands on the machine every day, i.e., line workers (over engineers who work in cooled rooms or laboratories and over consultants who do not know the work place very well).

4.2.2. Multi-machine manning worker (MMW)

The wide range of the number of machines/processes a worker can handle shows that the multi-machine manning system can work even in the small and mid-size manufacturing.

A worker can handle as many as ten machines at once. This supposes that the tact time, that is the time a worker performs a specific action on a machine and the order of the different successive actions to be taken are well defined. This reminds of the contents of standard operations. There is a clear indication that if a company switches to JIT methods, its work force can handle many machines at once. It is interesting to note that the minimal maximum number of processes/ machines by worker is two.

The forces of changes toward JIT have been detected in only 36 companies that have MMW. The five companies that clearly stated to use the single-machine manning system and the 88 that did not answer the question can all be considered as representing the forces that resist the change of the status quo (Figure 4–18).

4.2.3. Stocks and wip

A minority of 32.40% of the companies surveyed practice a thorough zero stock production. That is very suggestive as to the possibility for other small companies to do so and as to the directions toward which small companies may be bending. Those small companies are even ahead of some giant corporations like Toyota which has initiated the JIT system. In fact, Toyota keeps always a small inventory. This fact is confirmed by Zenzaburo Katayama, assistant manager at Toyota Motor's TQC Promotion Department:

> *People sometimes refer to the Toyota production system as a "non-stock systems. However, this is not correct. We always have some stock at hand, since we do require a certain inventory level in order to produce the necessary number of products at a give time …* [18]

The two distinct forces moving in opposite directions can well be observed at those companies that have stocks. Companies that run stocks worth almost a year represents the conventional methods and those for which the volume of stocks is equal to or less than 20 days can be said tending toward JIT and ZS productions.

Therefore, it may be said that there is two tendencies as regards the stock policies. Because before JIT, companies were run on a stock basis, it can be stated that now some companies are switching to the non-stock production and that even a few of them have already achieved that objective.

JIT considers also WIP inventories to be a kind of waste that should be eliminated. Concerning WIP, the tendency toward NS production is clear.

The presence of the two opposite and conflicting forces can is revealed by the following data. 68 companies have both stocks of finished goods and WIP and 15 companies have neither WIP nor stocks of any kinds. 38 companies have no stocks of finished goods while 26 do not have any WIP. 10 companies have stocks of finished goods but no WIP. And 18 companies have no stocks of finished goods but they do have WIP.

---

18   M. Imai, *Kaizen. The key to Japan's competitive success*, 1991, p.91

The number of companies whose minimal stock is less than 10 days represents 21.51% while for the maximal quantity of stocks, it falls to 13.73%. For WIP, the figures stand at 25% and 22.29% respectively for the same number of 10 days. That indicates a bias toward JIT production.

4.2.4. Mixed Model Production (MP) and lead time

The tendency toward the mixed-model production (MP) is evident. In fact, according to the surrey findings 92 companies out of 129 use the mix production or order-made production and the forces of inertia representing the conventional manufacturing of a single production on one processing line has a frequency of five (the frequency represents the number of companies). And the production lead time is not long for 42.11% of the companies for which less than a week is needed to complete the production of their products. Traditional methods are clearly represented in two companies for which the minimal production lead time is 150 days.

In a word, it can be said that concerning the mix production and the production lead time, the tendency toward JIT is clear and neat.

4.2.5. TPM

The total productive maintenance should be considered as being at its infant or burgeoning stage in the small and mid-size manufacturing: only six companies do use TPM and 35 companies do not even know anything about it. The six companies that practice TPM should be considers as pioneers of the new trends.

4.2.6. Supplier relationship

As for the supplier relation, the shift toward JIT can be sensed by the fact that you have companies that receive the ordered products on a daily basis with a delivery frequency of more than once a day in some cases. To that category, one has to add the companies who have weekly deliveries and for which the delivery time is equal to or less than two weeks. One might be inclined to think that the question missed, concerning the delivery time, one point. It did not ask to specify if the delivery type (daily, weekly or monthly and the number of times within each of them) concern each of the suppliers or that represents all the supplied parts in general.

Fortunately, the difficulty was overcome by the fact that each company had to specify the delivery time for ordered parts. And the delivery time that takes only hours or days means that the ordered product is delivered within hours or days.

4.3. Summary

At the first sight, it seems clear that JIT as a system of production seems quite unknown in the small and mid-size manufacturing. A deep look into the matter reveals however that some of its elements can already be detected in the small and mid-size manufacturing. Are those elements just penetrating the small and size enterprises? Have they taken roots or are they just taking roots in that manufacturing sector? Unfortunately neither the speed nor the extension/intension of the forces of changes toward JIT in the small and mid-size manufacturing can be determined at this level of research.

Part III JIT/Japanese management relationship

Chapter 5
Understanding JIT within Japanese management

Chapter 6
Understanding Japanese management from inside

Chapter 5 Understanding JIT within Japanese management

The analysis of the components of JIT will be done in three steps. First, in order to grasp the nature of JIT elements and their internal relationship, a tentative taxonomy of JIT constituents is going to be suggested. Next, as some elements of the JIT production system are shared with the Japanese management system, the relationship between the two systems will be investigated carefully.

At last, due to multiple, if not strong, ties between JIT and Japanese management, the question whether JIT can work in a different management setting as efficiently as it does in Japan will be raised and brought to the discussion table.

5.1. Understanding JIT from inside: classification of JIT features

Decades of research devoted to the JIT system have led to distinguish two main kinds of JIT elements: JIT industrial engineering techniques and Japanese-management-related features of JIT

5.1.1. JIT industrial engineering elements

It is a question of JIT elements belonging/related to the field of industrial engineering. They can be divided into two groups. There are on the one hand pure industrial engineering methods and on the other, industrial engineering elements that interact with the human beings, i.e. operators.

5.1.1.1. JIT pure engineering elements

In that category, one has to find techniques that are universally valid like laws of physics or mathematics. They have no close relationship with the social, cultural, economic or managerial environment in which were or are discovered for the first time. Those elements can therefore be applied anywhere in the world and yield the same results. Following are elements of the JIT production system I have identified as belonging to this group.

1) Quick set-up;
2) Autonomation (poke yoke or automatic stopping devices, full-work system);

3) Shop floor reduction;

4) Production of different kinds of items on the same line;

5) Breaking barriers between processes or sections/departments (in the physical organization of the shop floor) so as to realize a one-at-a-time processing line;

6) Arrangement of processes and machines in order to create a flow of products;

7) U-formed processing line (in the physical organization of the shop floor).

Those engineering elements constitute what I would like to call the "technical side" of JIT or just simply "technical JIT". Applying them would unavoidably contribute to the reduction of cost, production lead time, defective parts (work), overproduction of work-in-process inventories and work-force.

Are those elements related to each others? Though it seems difficult to answer the question by just yes or no, four observations can however be made.

First, the quick setup should at all costs be considered as having an order of precedence over the other elements, especially while dealing with the introduction of JIT. In order not to be trapped by the famous economic lot size, the shortening of the changeover time should be the first thing to realize before starting, for example, to produce different parts/items on the same line. Furthermore, it is meaningless for machines to stop themselves in cases of malfunctioning or slight errors if changeovers take many hours. The wise option would be of course to correct errors by rework as is the case in the conventional manufacturing system.

Second, though the reduction of the shop floor space shrinks the production lead time by curtailing the transportation time, that would not be so significant if there is no flow of products and that the products, in order to be moved from a process to another have to go through twistingly complicated ways.[1]

Third, breaking barriers between processes makes sense only if the production of detectives is neither allowed nor tolerated.

Otherwise, barriers would be required because of the necessity to check the acceptability of each lot before it moves to the next process. Barriers would be in that case necessary as they would fulfill the role of inspection stations.

Fourth, the U-formed processing line would be impossible to realize if barriers between processes are not torn down and if machines and processes are not arranged in the sense of the flow of products.

It is worth pointing to the fact that both autonomation and barriers breaking feature each a technical and a managerial aspect. Therefore, they will be mentioned also in the group of JIT elements involving the worker's activities or operational techniques.

---

1   See R. J. Schonberger, *World class manufacturing: the lessons of simplicity applied,* N. Y.: The Free Press, 1936, p. 1.13

5.1.1.2. Worker's operations/activities as JIT elements

The worker's operations can constitute and they do constitute some JIT elements. In other words, you have JIT techniques that interact with the person of the worker. Therefore their application and realization as well as their success depend also on the human factor. If they are accepted by the work force, then they can work, otherwise they can not. In that group, belong the following JIT techniques or methods:

1) Multi-machine manning working system;
2) Standard operations (doing what should be done the way it should and within the cycle time);
3) QCC;
4) SS;
5) Continuous improvement.

The first two elements are closely associated and the core is the multi-machine handling system. The standard operations can be understood within the multi-machine manning system. In fact, the standard operations whose content is made up of the operation instruction sheet, the tact time and the cycle time can be seen as a method, a means for a) coordinating and harmonizing different actions or operations of a multi-machine manning worker and b) synchronizing them with those of other multi-machine handling operators working on the same production line.

I consider QCC and SS to be the best instances of improvement activities (see Chapter 2). And improvement is the backbone of JIT and one of its permanent and un-reachable goals.[2] In the context of the JIT system, standard operations are under continuous improvements and the number of machines an operator can handle varies continuously thanks to improvement activities.

5.1.2. Japanese management-related elements of JIT

In this classification are included elements or methods of JIT production that are either borrowed/imported directly from or conditioned by the Japanese management. In that category a tentative effort was made to include the following techniques.

1) Breaking barriers between processes from the point of view the paper work and work functions.
2) Autonomation (decision by worker to stop the line);
3) Job rotation;
4) On-the-job training.

---

2 According to M. Nemoto, *Total quality control for management: strategies and techniques from Toyota and Toyoda Gosei,* 1987, there is no limit to improvements

Let me say a word about the elimination of barriers and autonomation. Breaking barriers means here eliminating the paper work that has to be done before the move of products from a station or process to another or from a section to another, say, from the forging process to the stamping process.

One may wonder why the human-related elements are not included as a sub-group in the group of management features of JIT.

Let me discuss first the case of multi-machine manning system. The multi-machine handling seems for us, to be too technical to be classified in the category of management-related features of JIT. I rather think of it as a set of technical actions or motion or behavior, requiring technical skills which do not have much in common with the pure management features. I admit, as I will show it later, the strong influence or the impact of the Japanese management system on the multi-machine working system because of similarities of situations one finds in both the Japanese management and the JIT system but I feel inclined to sustain that multi-machine remains a technique of industrial engineering.

If one thinks however that the multi-machine manning system and the standard operations they involve should be considered as management features of JIT, there is no fundamental objection against that classification. I think however that the approach proposed here reflects better the manufacturing reality of the Japanese factory.

The same question may be also raised about QCC, SS and improvement. Are they not elements of the Japanese management-related elements of JIT? QCC and SS and the notion of improvement they imply may be thought of as management features that JIT has adopted. I think that would be an error of perception. One should remember that QCC, for example, did not proceed or develop from the small groups which are characteristics of the Japanese management. They have their origins in the famous zero defects of

NASA and the quality control ideas introduced in Japan by Deming.[3]

Suggestions for improvement are closely related to QCC and can be seen as an emanation of QCC. The main difference between the two elements is that SS may involve and involves either an individual or a group of individuals while QCC is always a matter of an organized group or a team.

I have tried to put in the category of management-related features of JIT only the "raw" features of Japanese management. I mean by raw features the management characteristics that are found unchanged in JIT (e.g. job rotation). Those features are found not only in the factory management but in any kind of Japanese companies regardless of the type of industry the companies belong to.

---

3   See M. *Imai, Kaizen. The key to Japan's competitive success,* 1991, p. xxii

Those pure management features need no more explanation as they have been dealt with earlier.

With regard to what has been just stated, a word of explanation about the reasons that led to include autonomation among management features of JIT may not be a waste of time or that of energy. At the first contact, it is sure that autonomation sounds too technical. That is true when it refers only to machines and processes.

But then applied to the person of the worker, it loses its pure technical resonance. An autonomous worker refers only to an empowered or independently responsible worker and such a worker is not only the one confined in the production shop floor and who deals mainly with machines. The Japanese office worker is also very autonomous because he is given the power to perform many duties that are of the sphere of the management authorities in different management/cultural settings. Take for example the simple case of student's academic record transcripts. Both in the Congo and in Japan, they bear the stamp and/or signature of the dean. The main difference is that in the Congo the dean should sign himself while in Japan the dean's name is stamped by a clerk.

5.1.3. Kanban and kanban system

One should have realized that the kanban system was not classified. The truth is that I did not known how to deal with it since it since appears to be so special and so an important element of JIT. It did not seem to fit into any one of the categories I have constituted. It covers all of them and at the same time goes beyond them.

I have just avowed and incapacity to classify that important element of JIT though the least that can tentatively be put forward is that the kanban system is a subsystem of JIT. In fact, the kanban system is so vital and so important for JIT, especially for Toyota JIT that the latter is known also as the kanban system. Some JIT specialists assert that JIT can exist without the kanban system but the kanban system is meaningless without the JIT system. Others think that the kanban system is a transitional step of the JIT system and that JIT in its final stage will not need the kanban system[4] any more. In the light of this, one understands why for many people, the JIT and kanban systems refer to the same reality. That explains why my classification seems to have rightly failed to absorb the kanban system. Classifying it might have left the strange impression that JIT is a sub-system of JIT itself!

---

4   See I. Shinohara, *NPS (New Production System): JIT crossing industry boundaries,* 1988

5.1.4. Summary and conclusion: groups' interconnection

This section has consisted in analyzing closely, not in describing as I did earlier, the techniques that make up the production system called JIT. In fact, JIT elements have been scrutinized in order to determine their respective nature. According to their nature or specificity, JIT components have been tentatively grouped into a category or another. In a word, this is a tentative effort to understand JIT from inside by examining the nature of its elements and the relationship that they may have with each others.

I think that grasping the nature of the JIT components is like understanding the individuals who belong to a community. That approach helps in dealing with the community one wants to know about. In the similar manner, the comprehension of the nature of the JIT elements (and if possible their internal relationship) should prove an efficient way of 1) grasping JIT as a system itself and 2) examining the possibility of its transfer in a different environment.

For a number of observers, JIT may look only like a pure production method having nothing to do with the surrounding environment. One should, however, keep present in mind the fact that JIT was born and developed neither at a technical research center nor in an engineering department of some university. It took form on a shop floor. And in the workshop, you have the work force and the management as the most important role players. In fact, the work force performs its job within and through the company-defined management framework. That is why the work hypothesis has been that JIT as a production system draws many of its elements from mostly three sources: industrial engineering, work force (worker's operations', and (Japanese) company management. I hope to have succeeded in showing that.

I thought however that JIT elements related to workers' operations can be included in the group of engineering related elements. Therefore, for the sake of simplicity and for the reason just stated above, JIT elements have been grouped into two main categories: industrial engineering and management elements, the former being divided in its turn into two sub-groups.

The different classes of JIT elements are not independent. They are part of the same reality, i.e., JIT and they are, on the contrary, closely related to each others.

In fact, one should have realized that the job rotation and the OJT which have both been classified as management-related features are crucial factors in transforming the line workers into multi-machine manning operators.

Closely related to the multi-machine manning system is the U-shaped processing line that has been mentioned as being an engineering element of JIT. The U-form line becomes useful and effective if the work force accepts to perform many operations simultaneously. i.e., they take advantage of that kind of shop floor layout to performs a variety of tasks and operate many processes at the same time. Therefore its success depends also on that of the multi-machine manning acceptability by work force.

The U-formed line should be viewed as another technical tool of making the multi-machine manning system more efficient thanks to the flexibility it offers: it helps increase or decrease the number of processes or machines an operator can simultaneously handle.

Besides, it can facilitate the checking or recording of the processing lead time of each item because of the fact that the starting and final points may be at the same position. If the work force resists becoming multi-machine handlers, the U-like like would play only the role of a technical ornamentation.

Let me take, as a last case, the elimination of barriers. Breaking management-related barriers makes obsolete physical barriers of the shop floor and they can be eliminated also. At the same time, tearing down physical barriers increases the flexibility of the multi-machine manning operators.

One should have noticed that Japanese management has created and even transformed itself into an environment ideal for JIT and is prepared to transfer some of its elements to JIT in order to sustain its development, its effectiveness and efficiency. In fact, some elements of the Japanese management have become integral part of JIT. On the other hand, the Japanese work-force has accepted to work in such an environment. That is to say those management-related elements of JIT, directly or indirectly, imply the involvement and the participation of the workers because the latter ones are the principal actors.

Japanese management and JIT seem so close to each other that the following two questions are worth asking: To what extent is JIT influenced by the Japanese management? What are the nature of relationship between JIT and Japanese management?

5.2. Understanding JIT within the Japanese management system

This section will try to understand JIT within the framework of the Japanese management system, i.e., it is an endeavor to show the impact of management features on JIT.[5] It is question of JIT elements that require a management environment capable of creating conditions favorable for their development and sustained development, i.e., a surrounding environment allowing them to be effective and work efficiently.

Two points will be examined here. The first effort will consist in trying to trace the impact of the Japanese management on JIT. Next, I will try to put forward

5   I have partially dealt with this matter in L. Kupanhy, "Understanding JIT within the framework of the Japanese management system', 1991

the new concept of the *"prefiguration"* of the Japanese management and JIT in the small and mid-size manufacturing. There is somewhat an objection thus to considerations about the Japanese companies based on the concept of dualism.[6]

5.2.1. JIT and management relationship in big corporations

In the preceding section (5.1.), I brought to the light Japanese management features that can be traced in JIT. It was, in other terms, a kind of detection or identification of the presence of management features that have penetrated the JIT system. This section focuses rather on the impact of the Japanese management on JIT. Looking at the matter from a different observation point, it can be said that this is an attempt of understanding JIT within its immediate surrounding environment i.e., Japanese management system.

Let me make it clear from the beginning that this is not another explanation of either the Japanese system or of the JIT system. Features of both systems will be recalled so far as interact.

Only will be emphasized the influence specific features of a system has on those of the other system. Japanese management elements not dealt with in detail in chapter one but contributing to the development of JIT might however be entitled to a word of explanation.

(1) Autonomation and decision process

The Japanese bottom-up decision process involves so many people that it is also known as the decision by consensus. And the fact that a decision itself may be initiated/suggested at the lowest levels of the company management is understandable only in the context of group management and can be traced in JIT where the decision power in the shop floor is under the entire responsibility of the worker or the machine to which, in Ohno's terms, "a human intelligence has been given". Both an autonomous worker and/or an autonomous machine can stop the entire processing line whenever he/it deems it necessary.

This is something unheard of in the conventional manufacturing system where stopping or shutting down the entire processing line is of the strict sphere of the management competence and responsibility. Usually such a decision is made by the highest management authority of the factory. That is to say that in such an environment, it would be difficult for JIT to feature the characteristic of autonomation, especially as applied to the human being.

---

6   I have in mind R. Clark, *The Japanese company,* 1987; and T. Nishiguchi, "Strategic dualism: an alternative in industrial societies", Ph.D. dissertation, Nuffield College, Oxford University, 1989 cited by J. P. Womack, D.T Jones and D Roos, *The machine that changed the world. The story of lean production,* NY: HarperPerennial, 1991

(2) Multi-machine manning and the recruitment system

The recruitment system of the Japanese company lets in only young people (15–22 years old). Young people are very flexible. Second, the recruitment as stated in chapter one is not based on special skills of the new employees. The capability to learn as certified by aptitude tests or examinations for recruitment, the candidate's potentials, is one of the most emphasized points. Consequently, the Japanese company recruits young employees who are, mentally speaking, pre-disposed to undergo the training at the work place and to acquire skills the company needs and that will make them useful at the company. On the shop floor the training will initiate them to handle different types of machines.

The Japanese company takes a double advantage offered by the flexibility and the potentials of the new employee and transforms him without any resistance into a multi-machine worker. From the first days he enters a company his spirit and attention are oriented on purpose toward the multi-machine function working system. The new employee is told he knows nothing and he is conscious of that reality. He knows also that he was not recruited for a special job requiring special skills. So he feels ready to accept easily the company philosophy and its working style.

(3) Multi-machine workers, job rotation and OJT

Through job rotation, another characteristic of the Japanese management, the Japanese worker will have to go from one kind of job to a different type of work and will develop many skills through that job rotation and thanks to the on-the-job training system. JIT companies use the job rotation and OJT to create multi-skilled or multi-machine manning operators on the shop floor Operators during their working lifetime are continuously being moved from process to process, from a group of machines to another different group of machines, even from a section to another. They never resist being rotated because this is not done only on the shop floor. All those who work for the Japanese company, management included, expect to be rotated. And changing work places inside a plant necessitates undergoing a hand-on-the-machine training. One can now understand why the JIT systems has succeeded and transformed the work force into a multi-machine manning labor force.[7]

(4) Multi-machine worker, turnover and re-organization

The total absence of the labor market that results in the low turnover can be seen as one of the decisive factors that lead/constrain the Japanese worker to

---

7   It is worth noting the first attempts at Toyota were fiercely resisted with the workers striking.

become not a specialist of a specific field but a generalist specialized in his company's affairs.

The narrowly focused specialization has no advantage and there are no incentives for such an effort since no one will be sought after because of his special skills. There are no markets for special skills.

Therefore the specialized skills almost can not sell and are not marketable in Japan. Besides that, the Japanese worker knows he will not change companies and that he will not be seeking a position that requires a highly specialized knowledge or skill.

That is why the Japanese worker, particularly, the one on the production line can accept without resistance to become a multi-machine worker. He has much at stakes to do so because he is rated according to his ability to master as many techniques as possible.

Furthermore, he knows that mastering many skills and techniques required by his work place increases his chances of being promoted or assuming, in a competent way, responsibility positions sooner or later. All the other things being equal, the one who is broadly more capable has a competitive edge over the others as regards the promotion.

The absence of the mobility of the work force and the seniority system of pay and the multi-machine manning can help restructure a company by eliminating some posts and making the organization look flat. As matter of fact, Toyota eliminated, in its plants, the middle class management posts and ranks without any fear of grumbling, protest or strike or protest on behalf of the workers because 1) those people who have held those posts did not suffer any loss of material advantages since salaries are mainly based on the nenko system and not depending on a type of function or special post or ability; 2) being multi-skilled, they would start doing other kinds of job; 3) a title is mostly recognized within a company, and when one happens to change companies, he will start in all probabilities at a lower stage if not from the bottom stage. So the elimination of the posts and the titles does affect so much the one who is going to stay with the company until the mandatory retirement age.

(5) Elimination of barriers

Figure 5–1 Flow of information channels: American vs. Japanese company

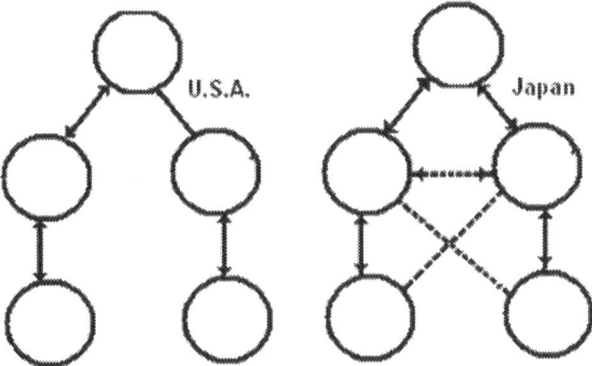

Source: Adapted from Kagono, Nonaka, Sakakibara and Okumura, 1985, pp. 106–107

In the American company like in its Congolese counterpart, the information flows vertically from the top to the bottom (instructions, decisions) and from the bottom to the top (reports). This seems in line with the sacrosanct business administration principle according to which a subordinate should report to one and only one chief. In that way there are no horizontal or diagonal exchanges of information.

In the Japanese company there are almost no barriers for the circulation of information. The information flows almost in all directions: vertically, horizontal (laterally) and diagonally. The information channels of the Japanese companies are made at will porous so that they seems to be leaking permanently

In such a context, breaking barriers on the shop floor poses no problems. And Toyota went as far as suppressing the middle management posts in the production department. Toyota did not face any avowed resistance because of the leaking nature of the information channels. In such a situation, formal barriers do not look necessary. Furthermore the reasoning of the Japanese worker seems to go as follows: if the elimination of information barriers makes the information flow faster and if that is beneficial to the company prosperity, it is also beneficial for the company employees.

(6) QCC and group management

Small groups are the basis of the Japanese management organization. One of the dynamic forces underlying the strength of the Japanese company is its group orientation. The organization of the Japanese management can be viewed as that of groups of groups.[8]

On the shop floor, small groups are known as QCC. The QCC movement has its origin in the manufacturing where the notion of quality control was introduced by Dr. Deming in the early 1950s.[9] The movement found a fertile field characterized by the presence of the already existing small groups. That may explain why QCCs are so popular and so successful in the Japanese JIT factory.

Nowadays QCCs have spread beyond the shop floor and have penetrated deeply the organization of the Japanese company as a whole. The phrase "company-wide quality control", known also as "total quality control", is used to refer to QCC structure that covers all the levels of the company organization.[10] Besides that, QCCs have crossed the manufacturing industry borders and are widely applied even in companies of the service industry. At the beginning QCCs might have meant a small group in the department of engineering or production but now it refers to any small group whose objective is to reduce the cost and improve the quality of product, service, work, etc.

(7) QCC, SS, competition and cooperation between small groups

Like the other small groups that characterize the Japanese management organization, different QCCs of a JIT company shop floor cooperate and compete with each others. Prizes are annually awarded to those QCCs that have the best performance and or have contributed the most to improvements. This is a proof that QCCs are stimulated through competition. The collaboration can be witnessed in the fact that good ideas for improvement made by a QCC will be shared with the other competing groups and adopted factory- or company -wide.

Within each small group, one will find the same structure of cooperation and competition. By the fact that a group has only one leader implies implicitly but directly that the competition is sometime always present among individuals of any organized group since in a way or another, the group members have to compete for the leadership position. The collaboration within a typical Japanese group can be found in the way the group performs its job. Though the group's task may be divided and assigned to its individual members, if one cannot finish his part of the job within the prescribed time, say the cycle time, other members of his group he competes with will come and help finish the work within the defined limit of

---

8   See Sh. Uemuraura, "The Japanese way of management: its characteristics, current practices, and future perspectives" 1989

9   M. Imai, *Kaizen The key to Japan's competitive success*, 1991, p. xxii

10   lbid., p. xxv

time. And when one is away, his job will be redistributed amongst those who are present. That is to say the cooperation exists at the individual level too.

SS is the structure devised by JIT in order to collect ideas and suggestions for improvement by Individuals. It is the structure that stimulates individuals to compete by conferring prizes for best suggestions to groups and/or to individuals with a certain number of accepted suggestions per year.

Suggestions can be authored not only by an individual but also by a group. And the proof that group members collaborate is given by the fact a group can submit a suggestion for improvement.

The American management organization that does not rely on the notion of group has proved to Mazda of America that QCC and SS can not easily and successfully be implemented.[11]

(8) QCC, SS, lifetime employment and seniority system

The job security or lifetime employment, the nenko system or the wage and promotion system based on seniority coupled with the promotion from within are important features of the Japanese management system. In such an environment where one's working lifetime is devoted to and spent in only one company, the company becomes in the mind of the worker, his company. Therefore the Japanese worker feels more inclined to search for possible means to improve his work environment which is also his own environment.

By the way, all my Japanese friends and acquaintances who work for private companies know well when the work starts and when the company officially closes but are not sure as to what time they will stop working and leave the workplace. They always say something like this: "In principle, the work ends (it means the company closes) at 17:30 but we usually leave the work place about 20:00 or 21:00 (that is the time lamps in the office may be turned off). It depends."[12]

This aspect sheds some light on the success of improvement activities in JIT companies in Japan. Further more, the question (often asked) whether QCC activities should take place during the working hours or after work seems to have melted and lost its real substance in the Japanese environment. QCC activities can be held on company time or after the work ends: that does not matter so much in the context of the Japanese management system.

In addition, because the Japanese worker considers the company he works in to be his thing as opposed to the Congolese worker or the American who lends/sells his services and skills to the shareholder's company, the former is more inclined to

---

11   See J. J.Fucini and S. Fucini, *Working for the Japanese. Inside Mazda's American auto plant,* 1990

12   This does not imply that overtime is recognized and paid.

search for possible ways and means to improve his productivity and the productivity of his work place. He is ready to devote his time, even his free time to QCC activities and SS.

(9) Improvement suggestions and bottom-up decision process

Improvements suggested by the line workers can be given a serious consideration to only in the framework of the Japanese management where ideas concerning decisions relating to the daily work routine or work execution can originate at a lower level of management. In the context of the Taylor's scientific management which distinguishes between the decision conceiver, evaluator and maker on the one hand, and those who have to execute the decision on the other hand, an idea for improvement originating at the lower level of the company is something nobody thinks of.

Proposing such an idea would mean an intrusion into the sphere of the management competencies by someone who is not entitled to do so and the proposition would not be given a minute of consideration. The suggestion will definitely end up slipping into the trash box.

Conversely, the worker would not lose his time thinking about improvement since it is not his job.

That reminds me of the old Japan when Matsushita was an employee at an electric company. He proposed an improvement for the socket of the bulb lamp. He was laughed at by his superior who did not waste his time looking at his design and rejected the idea. That led Matsushita to quit that company and set up based on his improvement (invention), his own company, internationally well known as Matsushita/Panasonic.[13]

(10) Improvement, cost reduction and bonus

The Japanese worker knows that if his company reduces the cost and earns a lot of money, this profit will have some direct effects on his annual income since he is going to share the company profits in the form of a bonus. And the volume of the bonus, in many companies is promotional to their financial performances. So in Japan where the company prosperity may mean a bigger bonus for everyone, but not only more dividends for shareholders, workers commit understandingly themselves to QCC and SS activities: overtime work very often means "sa-bisu" i.e., a voluntary service to the company free of charge for the latter. And there should be no surprise if some good suggestions for improvement have as their birth place a cafe or a bar where a small group of workers (QCC) usually drink

---

13   See K. Matsushita, *Quest for prosperity. The life of Japanese. Industrialist,* Kyoto: PHP, 1988

together and discuss at the same time the company affairs. That sheds some light on the fact that in some companies the number of suggestions reaches millions.

(11) Improvement activities and lifetime employment

One of the goals pursued by JIT through improvement is the reduction of the number of workers (hito-berashi) on the production line. In the context of the Japanese management, people do not fear proposing new ideas about improving the processes and machines even though the improvement would lead to the reduction of the number of workers. In fact, hito-berashi does not imply that people will be laid off.

People who are eliminated from the processing line will be assigned other jobs to perform.

For the same reason, a Japanese worker has no reason to resist and never resists the introduction of a new technology such the factory automation, robotization, computerization (FMS, CIM), etc. since the new technology threatens neither his job or nor that of his fellows workers.

(12) Stockless production and labor management relation

It has been established that the relationship between the labor union and the company is very productive and constructive, company growth-oriented since there are only company unions.

Some managers, including board members, were trade union leaders and current trade union leaders will be some of tomorrow's managers. The probability of strike is almost zero. In such a situation, JIT companies in Japan see no danger of running their plants without the safety stock. If the process is well controlled and defective-free, the need to stock disappears as there is no danger of strikes.

(13) Stockless production and supplier relationship

In the Japanese management environment, the supplier companies are not suppliers in the Western acceptation of the term.

The term refers rather to sub-contractors. It is capital to point to the fact that sub-contractors here almost mean permanent sub-contractors, i.e., a long-term partner or collaborating firm of a smaller scale. The sub-contractor is thus considered to be part of the buyer company which in the Japanese context is referred to as the parent company.[14] For its survival, the supplier sub-contractor depends almost exclusively upon its only parent company and the relationship is so firm

---

14 See M.J. Smitka, *Competitive ties. Subcontracting in the Japanese automobile industry,* 1991

that Sakai[15] thinks that Japanese supplier sub-contractor which is generally small or medium in size has no freedom.

In such a situation where the parent company maintains a strong hold on the suppliers, the parent company may rely on the latter to get what it wants and at the time it needs it. And one can understand why JIT has succeeded easily in Japan and is applied by almost all the giant manufacturing companies that depend on their suppliers which make more than 70% of the parts of the products made by their parent companies.

(14) Non-stock production, product life cycle and market competition

The Japanese market is a saturated market for many products such as automobiles, electric appliances, electronics apparatus, etc.

That is one thing and the other is that it is a market characterized by a very fierce competition. Following is a confirmation of my observation by a Japanese executive of a successful company of the manufacturing sector:

> *In Japan today there are more makers of civilian Industrial products than in any other country on earth, including the United States. And these companies—nine automobile makers and two heavy truck makers, more than one hundred machine tool makers, and over six hundred electronics companies, for example—are the survivors of the competitive struggle.[16]*

To keep a competitive edge, companies strive hard to improve the quality of their products. Putting in the market place improved products is the vital force that makes a company survive in Japan.

That means that the life cycle of products in Japan is very short. To extend the life of an article, companies strive to add new amenity features and features that make the product a little more attractive, more pleasant, more convenient, more efficient, etc.

In that context, a large stock is risky. Morita of Sony explains and illustrates at the same time the danger of possessing stocks in Japan:

> *The continued vigorous competition we have in Japan has also changed the way we look at how we work. In the past, it was Important to produce a large stock of a product at the lowest possible cost, but now the life cycle of our products is getting shorter and the cost higher, and if we build up huge*

---

15 See K. Sakai, "The feudal world of Japanese manufacturing", Harvard Business Review, Nov/Dec 1990, p. 38–51

16 A. Morita, *Made in Japan. Akio,* 1987, p. 204

inventories, we may find ourselves with a stock of outdated goods. The premium is now on how quickly and efficiently we can get a new product onto the assembly line. In the past we could run a particular model for year and half or two years; now we must change models in just half a year, and often sooner. Sometimes it seems like a great waste to put so much investment, so much sophisticated technology, and so many highly complex procedures into such a short cycle, but if we try to lengthen the product cycle and market a model longer by sticking to an old design, our competitors will be in the marketplace with a new model trying (and perhaps succeeding) to take the business away from us. (…)

I reminded my managers at a worldwide meeting in 1985 that our major competitor came onto the market in Japan with a small compact disc player only seven months after we introduced this new technology in 1984 with our small CD-5 player. In fact, the competing machine was even a bit smaller than ours. Initially, we had not produced enough stock of the new tiny CD player—It was an instant hit—before other small players began to appear on the market so we were short lust at the time we needed to have big stocks on hand. Fortunately, our customers who couldn't get the tiny model CD-5 bought out all of our more expensive models instead. So the story had a happy commercial ending for us.[17]

Summary and observations

One should have realized that the impact of the Japanese management on JIT is very strong. First, there are features that JIT has borrowed, unchanged, directly from the Japanese management.

Second, the influence of Japanese management does not consist in a simple correspondence or correlation between two features, say, a management characteristic on the one hand and a JIT feature on the other.

On the contrary, the impact consists mainly in the creation, by the Japanese management, of an environment ideal for the development of some JIT techniques. In fact I have shown that a JIT component is usually conditioned not by a management feature but by a set of management features.

The QCC and SS development is influenced by group management, competition and cooperation between groups, wage and promotion systems and the life time employment.

Improvement's favorable conditions are created by the bottom-up decision process, the bonus distribution system that does not exclude the line workers and lifetime employment.

---

17 Ibid., p.209–210

The multi-machine manning system can prosper in the Japanese company because it is sustained by the recruitment system, the job rotation and the on-the-job training.

The stockless policy which is also a JIT feature can be justified first by the cooperative relationship between management and the labor union that represents the workforce. To that management condition, one should add the Japanese industry environment characterized by the high competition on the market place and the strong relationship between the maker (Buyer Company) (who can be the parent company) and its sub-contractors or part supplier companies (who might happen to be the "child" companies). Makers and subcontractors make up a large family within which the cooperation is very strong.[18]

Only autonomation seems to be influenced by one management feature. That feature is the decision process (that can be left at the lower level of the company). It should however be understood that group management exerts a direct influence on autonomation and that there should be indirect influences. In fact, the decision process such as applied in the Japanese management is possible because of the life time employment. Therefore it can be inferred that lifetime employment has some indirect influence on autonomation. Due to management features' logical interconnection (see Chapter 1.1.2.), there should be more indirect influences.

Let me point out that JIT techniques are sustained each by many management features and that a management feature backs generally more than one JIT feature.

The last observation will concern the fact that among the JIT components profiting for their full development from the management-prepared environment some are JIT constituents (autonomation, QCC, SS, multi-machine manning) and others may be viewed as JIT pursued ultimate objectives or JIT basic philosophical elements (endless improvement, stockless production).

5.2.2. JIT and management prefiguration in the small and medium size enterprises

The work hypothesis here is that the Japanese big corporation management and production methods are prefigured in the small/mid-size manufacturing companies as is also their relationship.

I will examine the management prefiguration first and will discuss the case of JIT next.

---

18   See M.J. Smitka, *Competitive ties.* 1991

5.2.2.1. Prefiguration of the Japanese management in the small/mid-size manufacturing

Many scholars pretend that Japanese big and small/mid-size companies feature two quite different management systems. And the difference between JBCM and the small/mid-size company management (JSCM) has thus been described and/or referred to in terms of duality, dualism, stratification or classes.[19]

In fact, for many specialists of the Japanese company, the Japanese company management (JCM) =JBCM. Therefore they give no considerations at all to JSCM which is for them a management reality completely different from JCM/JBCM.

That sheds, I think, some light on the rigidity of the concepts they use in order to refer to the separation between the two management realities.

None of the explanations based on those concepts has satisfied me, i.e., none seems to reflect adequately the reality of the Japanese company management (JCM) in general. Theories based on those concepts suggest, in a way or another, a static situation of two distinct realities with nothing in common.

I have at least one convincing reason to think that JBCM and JSCM entertain dynamic relations inside JCM and that it is necessary to find a concept that incorporates that dynamism. In fact, it has been demonstrated thanks to the field research that the Japanese big and small/mid-size companies share many features. That means that the small and mid-size manufacturing company has some basic features of the big corporation. Therefore, I thought that what differentiates JSCM from JBCM might be the weight given to their common features.

The attempt consists in approaching JCM in its totality. *It is an effort to get an integrating view of JCM that does not exclude a category of corporations but that covers the notion of JCM so that it can apply to the big as well as to the small/mid-size manufacturing companies. The contention here is that JBCM and JSCM are both integrating parts of JCM.*

Their relation within JCM can be grasped and expressed adequately by the concept of prefiguration that we are tentatively putting forward.

Let me take JBCM as a model, the top of JCM. In such a situation, I think that JBCM can be regarded as the last stage of the development of JSCM. In other words, JSCM looks like prefiguring JBCM. In fact, a small company that grows steady in size will sport a management style that tends more and more toward JBCM. In the same way, a big corporation that is losing its importance because of some difficulties will have a management style shifting toward JSCM. All those swinging movement between JBCM and JSCM takes place within the management framework that I call JCM.

---

19 See R. Clark, The Japanese company, 1987; W.G. Ouchi, *Theory Z,* 1982; R. Benedict, *The chrysanthemum and the Sword. Patterns of Japanese culture.* Tokyo: Ch. E. Tuttle, 1954 (47th printing, 1992)

The notion of prefiguration as regards the theory of JCM can be understood from two different but complementary angles of observation: extension and intension. Extension and intension are the two faces of the same reality.

Extension of the prefiguration of JBCM

The extension of JBCM prefiguration in the small/mid-size manufacturing refers to the number of small and mid-size manufacturing companies that feature JBCM characteristics known in the academic world as JCM. The extension deals with the number of individual companies of small/mid-size manufacturing sector to which a feature of JBCM can or may apply.

The range of the extensive prefiguration can be expressed in terms of the percentage of the number of individuals to which a feature may apply.

A hundred percents would mean that the feature is common indistinctively to both big and small mid-size manufacturing companies. A zero percents would mean that the feature is not found at all in the small/mid-size manufacturing.

Taking the two extremities as points of reference, one can speak of the strength of the prefiguration, zero being the absence of JBCM prefiguration in the small/mid-size companies and a hundred percents expressing the full prefiguration, i.e., the fact that the feature is shared by both JBCM and JSCM.

The prefiguration intensity or level may be very weak, weak, medium, strong, or very strong. The notion of duality or dualism is to be situated at the level zero of the degree of JBCM prefiguration in the small/mid-size manufacturing companies.

The development or growth consists in increasing the strength of its prefiguration, i.e., a rise in the number of the small/mid-size companies in which prefigures a feature of JBCM. The regression is the opposite movement consisting in the increase of the number of companies that are getting rid of a JBCSM feature.

The growth or regression here can be grasped in a diachronic perspective. As the time passes, a feature or set of features may be the object of adoption or rejection by an increasing number of companies.

**Figure 5–2 Possible diachronic evolution of the extension of a
prefigured characteristic**

Number of companies

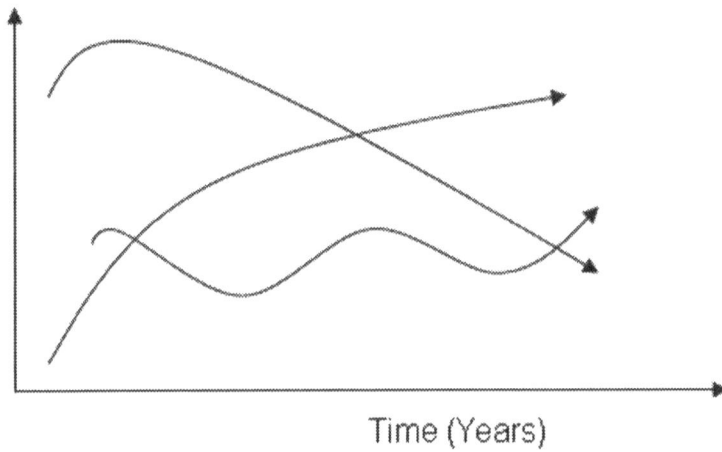

Time (Years)

The survey I conducted, like most surveys, can be seen as having checked the prefiguration of JBCM features from the extension point of view at a certain time. It tried to answer the following question: How many companies do feature JBCM i.e., what is the percentage of small/medium companies where JBCM feature are prefigured now? Table 2–3, (column "survey results", sub-column "frequency in %") summarizing the survey findings about management gives the prefiguration extension of JBCM features (except for management origin and turnover data) in the small and mid-size manufacturing. The prefiguration strength varies between 6.5%5 and 99.14%.

The prefiguration extension offers an advantage. It helps determine at a simple glance whether a JBCM characteristic is present (i.e., adopted or applied) or not in the world of small/mid-size companies. Furthermore, it shows the extent to which the feature is prefigured. The approach presents however some difficulties that put some limits to the material validity of conclusions that one is entitled to draw. It does not reveal the weight of features.

In fact, a feature may well belong to both JBCM and JSCM. But it may have different orders of importance in each category. Take the features age and skill as promotion elements: age may have more weight than skills in the big corporations while skill may be given an order of priority over age in the small/mid-size company. The extensive prefiguration tells that both the large and small companies possess the two features but do not indicate the order of importance of one relatively to the other in each of the two environments.

Intensional approach to JBCM features' prefiguration in JSCM

This can be viewed as the other side of the prefiguration theory. Here the accent is put on examining the weight or importance of each feature or its order of precedence inside JSCM in comparison to the role the very same feature play in JSCM. That is to say that I take once more as the Ideal reference the feature weight in the big corporation.

The guiding question is: what is the weight of such an element comparatively with the same element's weight in the big corporation?

The prefiguration intension tries to determine whether JSCM features prefigured in JSCM play the same role in both systems or not.

Figure 5–3 Weight or intension (importance) of a prefigured characteristic

Weight	0%	<100%	100%	>100%
Intension	Empty	Less important	Congruent	More important

Figure 5–3 can interpreted as a formal representation of the weight scale of prefigured features compared with their importance in the big corporations. An empty prefigured feature is a JBCM characteristic that does not play any role inside JSCM. It is in fact not part of JSCM. Close to empty features are those characteristics of JBCM that are also part of JSCM but whose role in the latter is completely insignificant. Such features can be said to have a nominal or symbolic prefiguration in JSCM. A feature may play exactly the same role and have the same weight in JBCM and JSCM. The prefiguration intension of such a feature may be referred to as congruent with JBCM. It means that there is a total congruence of importance of the feature in both the prefigured environment and in its environment i.e. in JSCM and JBCM respectively. Otherwise, it is not less or more important.

The growth for the intension of the prefigured features does not necessarily mean the increase of their weight or role in the prefigured environment. It rather consists in moving close to getting an order of precedence or importance similar to that of the same feature in JBCM. It is a shift toward the point of congruence. In other words, a heavier prefigured item grows by losing its weight while for a less important characteristic it implies getting more weight.

The weight of a prefigured management characteristic can be observed over a period of time inside one company in order to see whether it grows or not. Clark alluded to that when he said that as

Marumaru (the company he studied) grew in size, and its management style started moving close to that of big corporations:

During Marumaru's early days, when the company was a small firm in the timber business in a remote prefecture, there had been little or no possibility of offering employees automatic promotion. Most of the employees had been taken in from other firms, and not from schools or universities, so that their ages were not correlated with length of service or experience, as they would have been in a large and long-established firm. (…)

As the company grew, however, it became increasingly difficult to promote people according to ability. (…) An even more important reason was that in becoming a large company Marumaru had consciously adopted the personnel policies that befitted its new position in the society of industry. It made the utmost effort to recruit employees from school and university, and cut down on the number of recruits from other companies.[20]

A much more interesting and useful approach consists in measuring simultaneously the prefiguration extension and intension at a point in time or during a period of time. Checking the extension and the intension of a feature at a fixed time shows a static picture of the importance of the prefigured feature with regard to the number of companies it covers and the averaged weight it has. Considering the extension and the intension of a prefigured feature over a period of time would reveal the evolution and development of the feature concerning 1) its spatial expansion over time; 2) the evolution of the average weight given to the feature for the period considered. The importance of such research is that it helps to see the direction a prefigured feature is moving to.

The survey may leave the impression to have focused on the extension aspect of the prefiguration dealing thus mainly with the question, "Does this feature exist in your company?" and not with the complementary question, "If yes what is its order of importance?" It would be an error of perception to think so because the survey did not exclude that aspect completely and did not limit itself to checking only whether a JBCM feature existed or not in the small and mid-size manufacturing.

It is necessary to let it be known that JSCM elements covered by the survey can be grouped as follows:

1) There are elements that can not be investigated from the intension point of view of their prefiguration notion and for which only the extension results are given. Among such elements, I can mention the lowest management rank and the retirement age.

2) Investigating the intension of a characteristic implies its extension. Indicating the number of people for example, holding management posts according to their

---

20 R. Clark, *The Japanese company*, 1987, pp.117, 156 and 175

provenance (promoted, recruited, borrowed) shows the relative importance or weight of each source. At the same time, it shows the extent to which such features do exist.

3) There are elements for which the prefiguration extension reflects the importance (weight) of the feature. The number of companies people left for retirement, resignation, lay-off reasons, for instance reflects the weight of each of the reasons. The number of companies recruiting from school and in the labor market reveals the weight of the relative strength of each source. This is to say that the survey itself reflects the fact that extension and intension are the two faces of the same reality and that checking one may directly or indirectly shed the light on the other side.

Besides the survey explicitly checked the weight of some features such as bonus, labor union, strikes, etc.

Do you pay bonus? That question checks the extension of this management characteristic, since the answer will reveal the number of companies using that management characteristic. How many times a year do you pay the bonus? How many months of salary is the bonus worth? These questions deals with the intension of bonus as a Japanese management feature.

The question about the number of people who belong either to the company labor union or to the outside union examines clearly the intension of the labor union characteristics. Dealing with the number of non-unionized workers is the same as dealing with an aspect of the labor union intension.

How many times did you experience strikes or strike threatening? The question refers to its intension.

The weight of the following elements were not checked: age, level of education, length of service, ability, skill, special knowledge and their influence or importance in promotion, salary and allowances.

The importance of their prefiguration extension, in some cases, suggests the order of their importance, i.e., their intensive strength.

The intension aspect of management from the prefiguration point of view is the most difficult to deal with because it intends to quantify the quality which is per se not quantifiable. It is easier to determine the quality of a quantitative work but rather difficult to quantify the qualitative work. The former is easily accepted to the mind while the latter is not because quality means absence of discrete number, something not dividable while quantity means number and numbers are distinct one from another. The reality of numbers means discontinuity and quality means something indivisible and/or uncountable for which the volume or intensity is expressed by words such as "much", "few or fewer". But everybody knows that the heat is a qualitative reality but its intensity however can be determined by correct measure instrument, say, the thermometer. Likewise, the number of people in management posts and their respective origin have helped measure the relative importance of the different sources. The answer to the question: "how many people are there in management posts who came from such or such source?" helps measure quantitatively a quality.

Summary

Considering the weight of the common features, JSCM has been said to have the prefiguration of JBCM if it has to be accepted that features of both JSCM & JBCM differ mostly in weight according to the fact that they apply to big or small/medium size corporations.

The thesis has put forward the notion of prefiguration because of its adequacy (congruence with the reality of the Japanese company) and of the flexibility it offers in explaining how a company, as it grows, over time, in scale and importance, will give, little by little, more weight to those features specific to big corporations. It will not acquire overnight features it did not have at all before. The growth or development from small to large corporations is not revolutionary, implying the rejection of all older habits and acquisitions of new ones suddenly as implied in the notion of duality/dualism or that of or classes. The move is evolutionary, consisting in the growth and development of the potentiality of the company.

The difference between small/medium size companies and big corporations has often been explained in terms of duality of structure or stratification or class of industry. The notion of dualism reminds of the opposition, if not the struggle between the mind and the body in the Christian tradition and the conflicting difficulties them to be conciliated. The idea of classes recalls the Marxist/communist theory that has being fading out and has lost its luster since the end of the 1980s with the collapse of the Soviet Block. I do not subscribe to those conceptions. The prefiguration notion can however tolerate the stratification explanation if the latter is seen in a dynamic sense and not in a static acceptation of the word.

5.2.2.2. Emergence of the JIT prefiguration in the small and medium size manufacturing enterprises

Is JIT prefigured in the small and mid-size manufacturing companies? And if so, can it be inferred that the relationship between JBCM and JIT is also prefigured the JSCM? The issues raised by those questions will be discussed theoretically first and the conclusions arrived at will then be compared with the results of the survey.

From a strong extensional prefiguration of some JBCM features in the small and mid-size manufacturing it can be deduced that in all probabilities management-related elements of JIT are also prefigured in the small and mid-size enterprises.

The reasoning is logically sound but because of the danger of symbolic and/or empty prefiguration, that would have been risky in the sense that the reality might contradict the drawn conclusion. It is worth reminding that a symbolic prefiguration is an extensive prefiguration whose element weight is close to zero while the weight of an empty feature is zero. The presence (extension) of such elements even in the majority of small companies can hardly lead to creating a pre-environment ideal for the JIT prefiguration.

It is only from the intensive prefiguration point of view that one can infer that the increase in weight of JBCM features implies the increase of the probability of having the prefiguration of JIT features. What I mean is that one may deduce from the prefiguration extension of JBCM in the small manufacturing the prefiguration of JIT provided that the weight of prefigured JBCM is substantial.

The correlation that may be established between the weight of prefigured JBCM and JIT prefiguration constitutes an aposteriori proof that JIT is strongly influenced and conditioned by JBCM.

Concerning JIT it is important to remind the following points:

1) JIT was born in a big corporation and is thus an off-spring of JBCM. That is why it bears some inherited marks JBCM;

2) JIT history is relatively very young compared with that of JBCM/JCM;

3) JCM was born and is said to be entrenched in the Japanese society and culture.[21] I tried to show that JIT has some of its roots in the JBCM. Therefore, the environment of JIT and JBCM, though not unrelated to each others, are however different.

Due to differences of their respective history, birth place and environment, the two could not be approached the same way.

Concerning the presence of JBCM features in the small/medium-size companies, the survey dealt with and was based on the following question: what features of JBCM can be found in the small/medium-size corporations? Because of the old history of JCM and its ingraining into the Japanese culture/society, checking just the presence of its feature may be justified. I suppose that JBCM features that are not present in the small/medium-size manufacturing have less or no more chance of penetrating there. I was checking a state of affairs.

On the contrary, JIT being young and having a recent history, I think that the approach should be different. The guiding question is: to what extent is JIT penetrating the small and mid-size manufacturing? In other words, to what extent is the prefiguration of JIT taking place in the small and mid-size manufacturing

---

21 See K. Urabe "Innovation and the Japanese management system, *Innovation and management. International comparison,* 1988, pp. 3–25; Sh. Uemura, *Nihonteki keiei soshiki,* 1993

companies? The question of the JIT prefiguration is raised in terms of the emergence of its prefiguration.

In fact, JIT seems to be penetrating small and mid-size manufacturing, i.e., its prefiguration is taking form and emerging.

The concept of emergence has the advantage of indicating that JIT is penetrating and that its prefiguration is under way. The survey can be thought of as aiming to verify the emergence of JIT prefiguration in the small and mid-size manufacturing companies.

The survey dealt with two engineering techniques of JIT: the mix production (MP) and TPM because that question relating to those two JIT techniques could be understood by any one filling out the questionnaire (1 avoided too technical questions). Concerning MP, what is important is that featuring it means and implies that the quick setup is applied. And the quick changeover is the most important and the first step to undertake in order to smoothly switch to the JIT production system. Most companies produced many kinds of items on the same production line. The presence of TPM would imply that the waste of making detective work/product is explicitly recognized as an evil to eliminate. The survey checked mainly the worker's operations/activities that make up the human side of technical JIT. We are thinking here of the multi-machine manning, QCC and SS. Improvement activities and the minimal level of stocks that are pursued as JIT objectives also covered by the survey. They are all prefigured in the small and mid-size manufacturing.

The fact that the prefiguration of JIT is emerging and their strong extension prefiguration suggest what is already known, that JBCM features are prefigured in the small company; and more important that their prefiguration is neither nominal (or symbolic) nor empty. It means that JSCM features are, congruently speaking, close to those of JBCM.

Management-related features of JIT. Says job rotation, on the job training (even the off job training) were already investigated in the part of the surrey devoted to management.

In fact, many JBCM features that exert some influence on JIT are present in the small and mid-size manufacturing. Those features are age, seniority system, job rotation, good management/labor relation, bonus for everyone system, no lay-off policy, internal competition and cooperation, competition between companies, and so on.

Their intension should be substantial.

That is to say that the small and Japanese manufacturing prefigures also an environment relatively ideal for JIT implementation. Therefore, it is not a surprise to have realized that many features of JIT are already present in the small and mid-size manufacturing enterprises.

In the small and mid-size manufacturing, a parallelism may be established between their management strategies and production methods in the sense that small and mid-size manufacturing enterprises do not have all management features as they do not have all JIT features in their full extent.

5.2.2.3. Summary and conclusion

The prefiguration can said to hold at three levels. First, at the management level, JBCM is prefigured in the small and mid size manufacturing where it is referred to as JSCM. Second, at the operations level, some JIT techniques have already penetrated small and mid-size manufacturing enterprises enabling them thus to use JIT while they grow bigger. Those are mostly elements that create an environment ideal for JIT to work efficiently and effectively.

Table 5–1 Japanese style of management and production methods

	Management Style	Production Methods
Japanese company in general	JCM	Japanese production methods
Japanese big corporation	JBCM	JIT
Japanese S&M company	Prefigured JBCM or JSCM	JIT prefiguration (emergence)

Because of its recent history, I think of JIT as penetrating or being adopted by the small manufacturing. That is why we tentatively prefer speaking of the emergence of its prefiguration. Third, the relationship between JIT and management features seems prefigured in the small and mid-size manufacturing.

The survey may look not to have focused enough on the intensional prefiguration of JBCM in the small/'mid-size manufacturing companies.

That is why I dared not draw any a priori conclusions, concerning the status of JIT in the small manufacturing, based on a pure observation of the prefiguration of JBCM. I was however sure that if JBCM is prefigured in the small and mid-size manufacturing and if its intension is consistent then JIT is likely to be also prefigured there. The survey tries to check that conclusion which the collected data fortunately corroborate.

Due to close relations between JIT and Japanese management in the big corporation as well as at the level of their prefiguration in the small manufacturing, does it make much to sense to think that JIT can succeed in countries whose management style is quite different from Japan?

5.3. Can JIT work outside the scope of the Japanese management?

There are more and more companies, especially in the United States of America that have implemented JIT production methods. The success of JIT in the USA is frequently echoed by scholars in scientific publications, professional magazines or books.[22]

One remembers that JIT techniques have been analyzed into management-related, pure industrial engineering and worker-related engineering elements. A close look at the cases of JIT implementation in the US reveals what everybody should know by pure deduction: Are easily implemented JIT techniques of the pure engineering.

In fact, most companies that have implemented JIT outside of the Japanese setting have essentially succeeded in introducing those engineering elements. Many reported or publicized cases of JIT success consist in the realization of the quick setup, the reduction of the shop floor space, the realignment of processes or machines and their effects in terms of the reduction of the transportation and lead time, the end result being the reduction of work-in-process inventories.[23] That is a convincing proof as to the possibility of implementing the technical JIT in a context different from the Japanese environment.

It is worth noting that the other two sides of JIT are as important as its technical aspect. Besides, it has been shown that all the three sets or subgroups of JIT are closely related to each others. Based on those observations, it seems logical to think that JIT would not work the way it does in Japan if it is not implemented in its entirety, i.e., a JIT featuring all its three aspects. We remain convinced that a partial implementation of JIT cannot yield but partial results.

The point of view defended here is that JIT can yield the same fruits as it does in Japan only and only if it finds an environment similar to that created by JCM. In other words, it is difficult for JIT to succeed or yield the same fruits as it does in Japan if the environment is not particularly favorable for it.

---

22 See Grieco, Jr., P. L., W. G. Michael and J. W. Claunch, *Just-in-time purchasing: in pursuit of excellence,* Plantsville, Ct: PT Publication, 1988; R. J. Schonberger, *World class manufacturing,* 1986; M. Sepehri, "How kanban system is used in American Toyota Motor facility", *Industrial Engineering,* February 1985, pp. 50–5; M. Sepehri, "Car manufacturing joint venture tests feasibility of Toyota method in U.S.", *Industrial Engineering,* March 1986, pp. 34–41; M. Sepehri, "Manufacturing revitalization at Harley-Davidson Motor Co.", *Industrial Engineering,* August 1987, pp. 87–93; 6

23 See R. J. Schonberger, *World class manufacturing,* 1986, pp. 229–236; P. Johansen and K. J. Mc Guire, "A lesson in SMED with Shigeo Shingo", *Industrial Engineering,* October 1986, pp. 26–33

I am going to look at some difficulties that may hinder the implementation of the whole JIT in foreign countries. The problem in implementing the full JIT would proceed from JIT elements identified as relating to the worker's operations/ actions and to JCM Let me consider, for illustration, just a few of them.

5.3.1. Multi-machine manning system

Line operators would easily resist being trained into multi-machine workers if office workers are not multi-skilled. The line workers will believe in the job rotation which creates multi-skilled workers if this is felt as a company culture which applies to everybody. And the notion of multi-machine manning should be understood in the general framework of multi-function that must be spread company-wide. Otherwise, line operators would feel that they are being exploited or at least that they are the only ones of whom much is required is from. That would look unfair and they would not subscribe eagerly to the multi-machine manning system.

Besides, in a management environment where you do not have the notion of lifetime employment and where labor and management form two antagonist classes or groups, it would be surprising if not a miracle to have multi-machine workers. Various skills of the multi-machine manning workers are useful and positively evaluated only in the company the operator specializes in. Outside his company, the multi-skilled worker looks like a generalist, someone who knows many things superficially and nothing in depth. And there are no labor markets for such multi-skilled operators. In fact, the open job market is segmented and defined by different categories of skills and experience in a specialized field. In such a context, it would be difficult for people to become multi-machine or multi-skilled workers.

5.3.2. Standard operations

The standard operations are closely related to the multi-machine manning; they would not work if tasks are strictly defined by individual and type of work. The cooperation that is necessary for standard operations to work harmoniously would not be possible.

5.3.3. OJT and job rotation

I think that people would resist undergoing the on-job training and job rotations if they realize that the company would thank them at the rise of the slightest financial problems and that there are no markets for in-house-developed skills especially at the lowest level of the company. Furthermore, not only multi-machine manning skills have no markets but they are not marketable. Job offering ads always read: Welder, at least five years of experience and never multi-machine manning worker, experience five years or more.

Therefore, as far as the single specialization is highly marketable and that the lay-off policy remains the short cut to strengthening the company financial balance sheet, the idea of the multi-machine manning system would not be accepted easily by the work force.

5.3.4. Elimination of barriers

Barriers should be broken not only in the work shop but also in many areas of the company organization and/or structure.

Besides, operators would not in all probabilities subscribe to the multi-machine manning system in an environment where there is a clear cleavage between management and "hourly workers" of the American automobile companies, for example. The terms of their contract is negotiated every three years between the representatives of shareholders (top management) and the representatives of workers (trade union).

Vertical barriers should also be torn down within the company organization: able workers should be given the chance to climb the company ladder instead of having their future bound to the shop floor and the labor union.

5.3.5. Autonomation

Having autonomous workers on the shop floor means that management people have to a certain extent, agreed on sharing if not giving up some of their traditionally recognized decision powers in the field of the factory management. It means that they must trust line operators to whom they would have to transfer some of their powers and prerogatives. The idea of scientific management distinguishing clearly between decision makers and decision executioner should be softened if not abandoned. However, in the conventional manufacturing system, the notion of autonomation as applied to the workers is hardly accepted due to the division of work created by vertical barriers.

5.3.6. QCC, SS and improvement

QCC and SS may be established but they will be efficient and effective if everybody takes actively part in them and that the suggestions for improvement are taken seriously.[24]

Table 5–2 Comparison of statistics on SS: USA vs. Japan

	USA	Japan
No/100 employees	11	3235
Adoption	32%	87%
Participation	9%	72%
Average reward	$ 491.71	$ 2.50

Source: Robinson, R. G. and D. M. Schroeder, Training, continuous improvement and human relations: the U.S., TWI programs and the Japanese management style, California Management Review, 1993

There are some risks if QCC and SS are set up but can not bear any fruits. SS activities seem very expensive in the USA and very cheap in Japan. At the same time the contribution and participation in improvement is very low in the USA and very high in Japan (See Table 5–2).

Should QCC meetings be held during the working hours? The answer 'yes' implies that they are paid and supposes that they are not voluntary, i.e., everybody should take part in them. But if the participation in QCC is not voluntary, uninterested workers would be physically present but would not really participate. In that case, QCC may become financially a burden if their activities are held during working hours. If QCC meetings are held after work, they should be voluntary in a foreign environment where the company belongs to the shareholders and their representatives, i.e., management; and the time devoted to QCC activities after work should at any costs be considered overtime work. But how many would devote their precious free time to QCC activities even though it is considered overtime by the company? It was slated that suggestions for improvement are made either by individuals or by groups. In an individualistic society like the Western one there might be some doubt about group suggestions pouring steadily into the suggestions box. Besides, due to the high mobility of the work force and to the large and competitive labor market, having some good improvements at

---

24   See the failure case the JIT introduction at Mazda of America as reported by Fucini, J. J...and S. Fucini, *Working for the Japanese. Inside Mazda's American auto plant,* 1990. According to the authors, Japanese managers contrary to what they had stated during the training sessions, did not pay any serious attention to suggestions by American workers

one's credit increases the chance of securing a better job elsewhere and of being spotted or hunted. This is particularly true for engineers.[25]

The dedication to QCC, SS and continuous improvement by the worker is possible if he feels the company really cares for him and that he can benefit from the fruits of improvement in terms of financial rewards. The American typical company reserves enormous bonus for management and the worker would get nothing even if he gives the best of himself to make the company earn a huge profit. Besides, the American company is prone to lay off the worker in case of any financial problems in order to preserve the management jobs and the interest of shareholders. In such an environment, though companies may switch to JIT and set up QCC and SS, nobody can however expect them to be getting millions of suggestions for improvement as is the case in some Japanese companies in Japan.

5.4. Conclusion

JIT seems so rooted in its birth environment that it compels to think that it can not, in a foreign management context, work as efficiently as it does inside JCM. This means that introducing JIT in a new environment implies transferring also some JCM features, i.e., the creation of a Japanese-management-like environment. I feel thus the necessity of grasping in depth the JIT environment. I will then examine once more JCM in order to get a full understanding of its internal basic forces of cohesiveness.

---

25   See M. Imai, *Kaizen. The key to Japan's competitive success*, Singapore: McGraw-Hill, 1991

Chapter 6 Understanding Japanese Management From Inside

Japanese management will be understood in terms of its strategies/objectives and principles underlying its structure.

Figure 6–1 Some Japanese corporate strategies and gheir goals

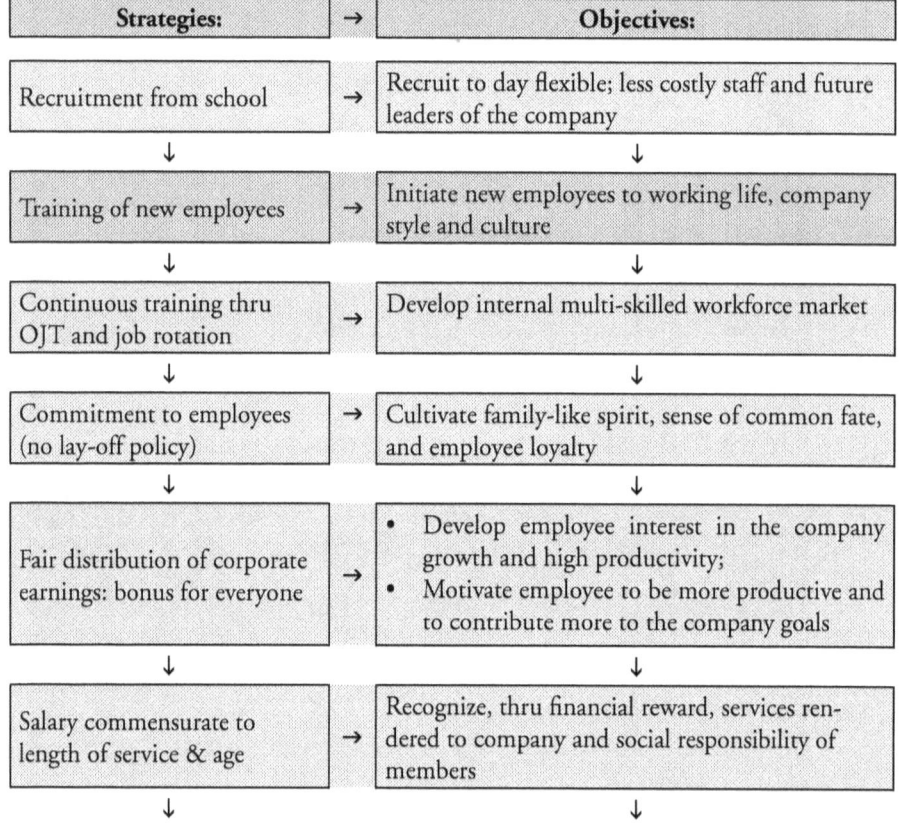

Strategies:	→	Objectives:
Recruitment from school	→	Recruit to day flexible; less costly staff and future leaders of the company
↓		↓
Training of new employees	→	Initiate new employees to working life, company style and culture
↓		↓
Continuous training thru OJT and job rotation	→	Develop internal multi-skilled workforce market
↓		↓
Commitment to employees (no lay-off policy)	→	Cultivate family-like spirit, sense of common fate, and employee loyalty
↓		↓
Fair distribution of corporate earnings: bonus for everyone	→	• Develop employee interest in the company growth and high productivity; • Motivate employee to be more productive and to contribute more to the company goals
↓		↓
Salary commensurate to length of service & age	→	Recognize, thru financial reward, services rendered to company and social responsibility of members
↓		↓

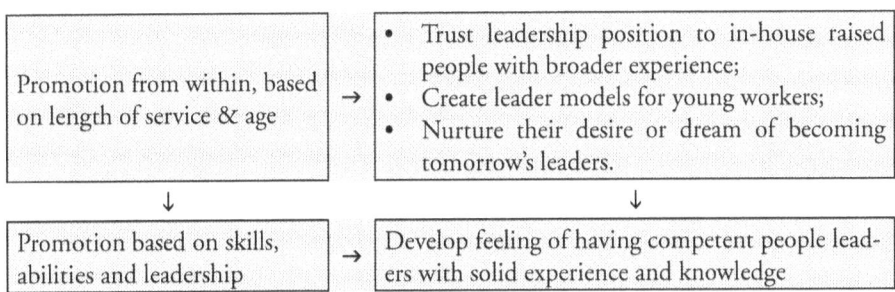

6.1. Japanese Management Strategies and their Objectives

The object of this section is to try to show that JCM features, besides their internal interconnection, may be viewed as playing inside the system the role of management strategies aiming at some objectives that should be thought of as JCM elements (see Figure 6–1).

Grasping JCM this way would help understand why one would like to introduce JCM. If one decides to pursue the same goals as the Japanese company, he may consider using the Japanese strategies or adapt them. The most important is to determine one's goals first: strategies are means to achieve those goals. In competing with the Japanese company, one should understand its goals and the means it uses to achieve those goals. The internal inter-connections of the JCM features dealt with in the first chapter show only the systemic aspect of JCM. No light is shed either on the goals pursued or on the means used to achieve those goals. By evaluating the goals and the way those goals are reached, one can understand the efficiency or effectiveness of a system.

It is worth noting that a strategy may achieve a multitude of goals as a goal may be pursued by setting up many strategies. I will review some management features that can be identified as strategies and state the goal(s) aimed at.

6.1.1. Recruitment

The recruitment in the primary work force market, i.e., the schools aims at long-term and short-term objectives. The short objective is to hire people who are not costly but potentially very promising, flexible, permeable and malleable and who can be indoctrinated and impregnated with the company culture, ideals and philosopher. Because they are not so costly in terms of remunerations, the extra money the company could save is used for their training

6.1.2. Training of the new employee

The training/indoctrination will bear their fruits in the long term when the company reins will be tomorrow passed over to those who are being hired today. The training new employees undergo aims at their initiation to the working life, the company style and culture.

6 1 3 Continuous training

The continuous training achieved thanks to job rotations, on-the-job training and even off-the-lob training contribute to the development of an internal work force market of multi-skilled workers who can perform different kinds of jobs and who know most aspects of their working place: the company produces its own experts.

The in-house training means developing internally the necessary market force. The Japanese company does so because it trusts those who work for it. The typical American company is known for relying heavily on the external work force marketplace in which it can recruit people with the needed kind of skills and the company lets employees go back into the labor marketplace as soon their skills are not needed anymore. On the other hand, those people with specific skills are prompt to move where their skills would be appreciated and sell at a better price.

6.1.4. Commitment to employees

The commitment on behalf of the company to its employees has as its end-results to cultivate the family-like spirit, the sense of common fate linking the company and its employees and to secure/condition thus the employees' loyalty.

6.1.5. Seniority-based payment system

The pay or employee's income's increase by age and length of service aims at developing in the person of the worker the feeling that the company recognizes the long and loyal services he has been rendering to the company. The financial compensation recognizes also the workers as having more social (family) responsibilities.

6.1.6. Seniority-based promotion

The promotion from within develops the sense of the company responsibility for senior workers and nurture the desire for younger people to remain in the same company in the hope that they may become tomorrow's leaders. For the younger workers this is a good and promising perspective and a motivation for not leaving the company while for senior workers promotion from within means the realization of their long-time dreams.

6.1.7. Skill-based promotion

Skills, capabilities and competence as complementary criteria for promotion to leadership posts develop within the organization, especially among younger employees the feeling that the company reins are in good hands. And younger people are thus stimulated to try to do their best in order to develop those leadership qualities which increase their chance of becoming, some day, company leaders. This skill-influenced promotion policy to which is added the length of service as a complementary criteria develops among the company members the sense of having leadership positions in the hands of competent people with the necessary experience. Therefore, younger competent people will patiently have to wait their turns.

6.1.8. Status equalization

The status equalization[1] concerns mainly the repartition of company profit and the promotion system.

6.1.8.1. Equal repartition of company earnings

The Japanese company looks fair as regards the repartition of the company profit. In fact, contrasting with its American counterpart which pays a bonus and offers stock options to management only, the Japanese company pays a bonus to everyone working for its growth, be he/she a manual worker or not.[2]

AII those who work for the Japanese company, managers and non-managers alike, benefit from the fruits the company bears by getting in the form of bonuses lump sums of money equivalent to many months of their salaries. Of course that system stimulates everybody to commit himself to the company success since the success of the latter contributes to his welfare. The strategy of providing the bonus payment to everyone should be considered as aiming at the best a worker can give to his company. The end result of this seen in the high productivity of the

---

1   I got the inspiration from K. Urabe, "Innovation and the Japanese management system." 1988, pp. 3–25; D. Waters, who speaks of the egalitarianism of the JCM, in his book, *21st century management: keeping ahead of the Japanese and Chinese.* N.Y.: 1991, pp. 47–48

2   Foreigners doing business in Japan see bonus from a different point of view: See J. N. Huddleston, Jr., *Gaijin kaisha. Running a foreign business in Japan,* 1990

Japanese worker, his dedication to the company and his willingness to improve the company output.

6.1.8.2. Equal opportunity for promotion

Another stimulant for the employee's productivity is the fact he feels that his fate and that of the company are linked together. His career path lies not through an outside labor union but within the company. He is conscious that being a union member constitutes a step among many others which are all directed toward some management posts: every worker feels he might become a manager[3] some day. In a word, there is not a clear line of demarcation between workers and managers. By the way, some companies require their university graduates to start their working life on the production line where they are union members.

6.1.9. Labor/Management relation

Because there is no sharp separation between management and labor, the latter feels to be really part of the company. The interest of the labor and that of the management converge completely.

Consequently, the management and the labor develop an interesting kind of relationships not based on conflict of interest, bargaining or opposition but on cooperation, collaboration and mutual understanding. There is no feeling, for one party of be exploited by the other.[4]

6.1.10. Decision making process and group management

Developing on behalf of the employee the sense of his being really involved in running the company is an important objective of the Japanese company's deci-

---

3   Urabe (1988, p.10–17) refers to that as status equalization but he fails to perceive the difference between strategies and the aimed results that are the goals

4   It is paradoxical to let it be known that sometimes it's those people in management position who complain to be exploited by the company they work at due to the fact that no overtime recognized when reach the management rank. See Clark, R., *The Japanese company*, 1987. I personally know someone who works for the biggest hotel in Wakayama and who complains that since he has been promoted a management position, he works harder and longer whereas his income went down because no overtime is paid any more. Besides, during a study tour to a K. Heavy Industry factory Kobe that makes trains and buses, to our question about possible cases of karoshi, we were told that they has never been a single case of "karoshi" (death due to overwork) amongst line workers although amongst managers, especially kacho, such cases do happen from time to time.

sion process. The Japanese employee feels he contributes to vital daily decisions which may lead to the success of the company or which may determine the course of the future success of the company. He feels he participates in the company management: he will be proud of that in case of success, will share the blame in case of the failure. He is a responsible worker. Another objective aimed at by associating everyone to the decisions the company takes is to get everybody's full commitment and dedication to the decision implementation.

6.1.11. Management independence from shareholders

Promotion from within means that management does not have outsiders as members of the board of directors (except some representatives of the group the company may be part of[5]).

Management feels independent and concentrates their energy on the company and the workers. They may sacrifice short-term interest and focus on long-term plans because they have no pressure from shareholders requiring a quick return on their investment. Therefore they don't feel under the threat of being removed from their posts if shareholders[6] are not satisfied.

6.2. Underlying management principles

The preceding section consisted in showing the strategic aspect of some JCM features and points out to their objectives. This is an investigation about the foundation of JCM from a strict management point of view. Till now, current management theories do not seem to pay so much attention to that aspect of JCM. I thought therefore that it needs some clarification. In fact, explanations abort the JCM foundation have been based mostly on cultural/anthropological approaches that show the entrenchment of JCM into the Japanese culture/society.

I think that my approach has the advantage of being new and of trying to understand the foundation of JCM as a management system from a different observation point, i.e., from inside instead of from outside. The originality of this approach is that it would shed a new light on that aspect of JCM. It pretends to be viewed not as superior but as complementary to socio-cultural approaches.

---

5   It is known that the main bank sometimes places its members in the board of the keiretsu group

6   It is known that shareholders in Japan have no power at all. This can be seen in the fact that the general meeting of shareholders is a pure formality that hardly takes over thirty minutes; the chief executive officer nominates his successor and is the most influential element in designating members of the board.

The analysis will consist in trying to find out some fundamental management principles on which is built the whole structure of the Japanese management. What is a principle?

6.2.1. Notion of principle

A principle is a source of actions, the reason for behaving in such a way rather than in that other. It is what gives some coherence to a way of behavior, to a code of conduct, i.e., the underlying force for actions. A strategy is a set of actions.

Ouchi links the success of a company with its philosophy.[7] In their explaining characteristics of successful companies, Pascale & Athos created the concept of superordinate goal[8] and identified its translation into reality at IBM, Matsushlta, and Delta. The superordinate goals of each of them differ however from those of the two others. In fact, the notion of company philosophy by Ouchi and that superordinate goal by Pascale & Athos seem identical and then interchangeable since they cover the same semantic field and thus overlap.

The notion of principle that is proposed here differs from the one of the philosophy of a company or its superordmate goal. My approach consists in a search for supra-company principles, i.e., common principles shared by the Japanese companies in general and that constitute the foundation on which is built the structure of JCM as a system. I do not feel concerned with individual company principles, philosophies or superordinate goals.

Besides, let it be known that the concept of superordinate goal proposed by Pascale & Athos is fundamentally different from the notion of principle. A goal, be it a superordinate goal, is an objective aimed at by an action of which the vital source is a principle. For us, a strategy is a set of actions aiming at the realization some objectives i.e., actions taken purposely in order to do or accomplish something.

It is a means for reaching a target. A principle is the basic reason why such actions are taken; it is their raison d'être.

A principle is much more pervasive and penetrates all the organization: It is invading, permanent. It is part of the organization environment like the air we breathe, to which we pay very little attention but of which the scarcity is strongly felt instantaneously and its importance brought immediately to the consciousness. The worker lives by and inside it like the surrounding air which is so vital to the luring being.

---

7   Ouchi, *W.G., Theory Z,* 1982

8   Pascale, R.T. & A.G. Athos, *The art of the Japanese management,* 1982

A strategy may last but it is not permanent: it is usually set up to solve a specific problem. It becomes obsolete once fixed objectives are reached unless similar targets are set again. If a principle changes, the structure built on it changes directions: it is a revolution.

It is necessary to distinguish between principles' effects and strategies' objectives. The application of a principle always produces some effects even if they are not sought for. Objectives are goals consciously set and aimed at by strategies.

6.2.2. Search for principles

Where to find basic management principles of the Japanese company? Cultural and social approaches[9] to the Japanese management system look outside the scope of the management. The approach used here is quite different in the sense that it is a tentative explaining of Japanese management from the managerial point of view. I have tried to find those principles within the JCM system. I therefore looked carefully among the JCM features and other management-related elements to see whether I could identify some that play that important role of principles.

Concerning the features of JCM, l wanted to know whether there may be some management characteristics which can be considered as the most important, i.e., the master-features or basic features and that are the basis of other features. Such features would be referred to as principles.

It may look surprising that features that were mentioned as strategies are now being reconsidered in order to identify them among them management principles. As a matter of fact, there may seem to be some confusion between principle and strategy. The reason is that a strategy may be elevated to the rank or status of a principle.[10] Take, for example, the case of an individual who sets as his goal a life full of prosperity. To achieve his goal, he decides to work hard.

Working is his strategy to achieve his goal. He may elevate that strategy to the level of a principle. He founds his life on the principle of work. All his actions will find their justification in the principle of work. Working would be his principle of life and his strategy for success. Likewise, I think that there may be some feature(s) that can qualify as management strategies as well as principle(s).

The basic principles are principles from which the whole internal structure of the Japanese management may be coherently explained or understood. They are mother-features that generate other features.

---

9   Abegglen finds the ultimate justification in the Japanese social organization, and so do Odaka, Hazama, and Uemura. Those cultural approaches are very sound. Mine is a tentative effort to explain Japanese management from within.

10   A principle can also be used as a strategy to achieve a goal.

In fact, I have identified some principles that I think, govern the structure of JCSI. I will try to group them into three categories: principles of 1) commitment, 2) interest precedence and 3) status equalization.

6.2.3. Statement of principles

6.2.3.1. Commitment principle

(1)   Commitment to all company employees and to reliance on them for the survival of the company.

That is the only principle that I have identified as belonging that category. In fact, after carefully screening JCM features and examining their possible nature/function, it appears to my mind that commitment to employees plays the role of a principle. Abegglen was the first scholar to have discovered the importance of that feature. But being mostly concerned with the comparison between the Japanese company and its American counterpart, he did not call it a principle but referred to it as the point of critical difference.[11]

The presence of that basic principle has of course some direct effects or implications which translate into reality by 1) carefully screening applications of possible employees and their initiation period during which they will be indoctrinated; 2) the continuous training programs for employees; 3) relying on one's work force as a whole and on each individual as a multi-skilled expert; 4) refraining from letting in mid-careers in order not to frustrate those already in the company and from laying off except in rare situations; 5) promoting people from within and rewarding financially those working for the company according to the length of service rendered to the company, etc. These direct effects may be referred to as first-level effects, i.e., effects emanating directly from the principle's broad application.

There are also effects of the second level resulting from the first-level effects: they manifest themselves in 1) the flexibility of the work force, 2) having as managers and executives people who grew responsible in the company and who thus know the company better; 3) creating the feeling of common fate, etc.

6.2.3.2.Interest precedence principles

(2)   Priority of company interests over workers, managers and shareholders

For the survival of the company, laying-off is possible, employees and managers may be transferred to subsidiaries, and dividends may be delayed or not be paid at all.

---

11   C. Abegglen, *The Japanese Factory,* 1979

(3)   *Priority of long-term over short-term interests*

For the survival of the company, short-term interests have to be sacrificed in favor of long term interests.

(4)   *Priority of more productive work force over others*

Usually, the Japanese companies reduce the number of new hirers first during hardships. Because of the current crisis due to the burst of the bubble economy, "Many companies, such as Toshiba Corp. and Nissan Motor Co., are now planning to decrease their personnel by not refilling vacancies."[12] Another category of less productive workers is made up of older work force nearing their mandatory retirement. Here forced retirement in the context of "the Japanese consensus" is going to take place in the form of early retirement or voluntary retirement.

The first three principles are being applied now as one can see it in the following cases reported by the Daily Yomiuri:

> *An Increased number of employers have recently stepped up streamlining efforts by encouraging their employees to retire and transferring them to affiliated companies. (..) Nippon Telegraph and Telephone Corp. announced at the end of August that it would encourage 10,000 of its employees to voluntarily quit. Tomoegawa Paper Co. announced on Friday that it would sell real estate in Tokyo and Kawasaki to cover retirement payments for voluntary retirees. Voluntary retirement temporarily increases the company's costs to pay retirement allowances, negatively affecting balance sheets in the short term.*[13]

(5)   *Priority of workers over managers' benefits*

This principle translates into reality by fact that when the times are getting hard for the company, the latter will first try to reduce management financial advantages by cutting their bonus and their salaries and other allowances before considering any measures that would hit workers.

(6)   *Priority of full-workers over others*

Part-time and seasonal workers are the first to be affected in cases of problems facing the company life.

(7)   *Priority of workers over interest of shareholders*

This principle application is seen in the fact that during critical periods, shareholders may be denied their dividends which are already by the Western standards very low.

---

12   *The Daily Yomiuri*, Sunday, October 3, 1993, p. 5

13   Ibid., p. 5

6.2.3.3.Status equalization principles

*(8) **Fair distribution of the company profit***

Fruits the company bears in the form of profit are distributed fairly among company employees and managers. Not only management but everybody is entitled to a bonus. The bonus is not the privilege reserved for managers as rewards on behalf the shareholders.

*(9) **No discrimination in the promotion***

The principle may be observed in action by the fact that the promotion which is from within does not in principle distinguish between blue and white workers. Any talented or capable person many reach high management posts. On-the-job training and job rotations undergone by everyone help develop the qualities and intellectual abilities that can play important roles during the promotion process. The egalitarianism is also seen in the fact that there is no sharp separation between management representing shareholders interest and the union which has as the main role to defend workers' interest only.

*(10) **Respect for the person of the worker***

The respect for the person of the worker is shown through the fact that not only managers but ordinary workers have a say in the running of the company. The workers participate by making any suggestion that can help better the company productivity. Lower levels of management are allowed to suggest decisions and even participate in the decision process. The contribution of everyone's mind to the edification of the company is what was referred to as group management which is possible only if everybody's opinion is given some weight in running the company.

6.2.4. Principles' general effects

There are many management features that can be considered: effects of those principles. I would like to have a closer look at only two of them, i.e., lifetime employment and low turnover.

6.2.4.1. Direct effect: lifetime employment

There are a lot of cries about changes in the Japanese management especially regarding the lifetime employment system. Since the beginning of this year (1993), there are many speculations about the explosion or the end of Japanese management.[14] This is understandable when one considers for example lifetime employment as a pillar or a principle of Japanese management.

---

14   See *Nikkei Business*, October 11, 1993, pp. 10–23

The study of the Japanese company has convinced me of the fact that lifetime employment is not a principle. It is the effect of precedence principles which do not exclude the possibility of laying-off. The commitment principle leaves open the possibility of laying-off in case there is no other way for the company to survive. For the survival of the company, even laid-off employees accept without any resentment their fate. Furthermore, Ouchi reports a case of someone who continued to work for his (ski) company and for no pay in order to help the company survive.[15] It implies that the exceptional cases of lay-off do not shake the commitment principle itself on which the concept of Japanese management is based. My opinion has been recently echoed by professor Uemura: Though there are changes in Japanese management, Japanese management does not change[16] at all. That means some effects may be affected by changes, but the principle being unshaken, the change is but superficial. Another echo is from Hajime Karatsu expressed in his recent book, "Japanese style of management does not die".[17]

Conversely, people are never laid-off in order to consolidate the financial balance sheet and in order to pay shareholder their dues Very often in the US, managers would lay-off employees in order to consolidate the balance sheet. That latter operation would push the company shares to go up in the stock market. As result the managers would get large bonuses due to the swelling values of the company stocks. Such a behavior is hardly defensible in the framework of JCM because there are no principles sustaining it.

6.2.4.2. Low turnover

The loyalty and the dedication of the Japanese worker to his company have two main sources. As a human being is by nature prone to trusting who he/she feels trusts him/her, the Japanese employee sensing the company's commitment and attachment to his person could not but commit himself to the company.

Besides, he has too much at stakes if he decides to leave his company: only his company recognizes at their exact value his multi-skills and polyvalence, his experience built within the same company and his company related expertise. Furthermore, outside doors are tightly closed so he feels blocked inside his company.

Should they be open, he is likely to be regarded a deserter and traitor. And there are many chances that he has to re-start his career at a lower stage if not from

---

15 W. G. Ouchi, *Theory Z*, 1982

16 The statement was made by Prof Uemura during his communication at a management workshop at Osaka City University

17 Karatsu, *Nihonteki ha shinazu*, 1993

the bottom of the ladder. In such a context, there is nothing a Japanese worker can do but devote oneself to his company and try to make his company a better place to work at and to live in.

6 2.5 Conclusion

As a matter of fact, I have realized that what is called management features of the Japanese company is a complex reality that needs to be considered from various points of observation in order to really get an integrated view of it. I have distinguished among JCM features those that qualify 'as strategies, objectives, principles principles' effects. Even some characteristics are playing the role of principles and strategies at the some time.

Not considering the strategic aspect of some of the management features would mean missing the opportunity to understand the relationship between those management features in a dynamic way: after defining the objectives, strategies are set up in order to move close to the fixed objectives. Features interconnection reveals only a state of affairs.

Besides, when introducing the Japanese management in a different environment, it is useful for one to distinguish between features that are objectives from those that are strategies. If one can see the difference between the categories of management features of the Japanese company, he/she may 1) adopt some strategies without knowing why and would realize later that the results obtained are not the desired ones; 2) find himself pursuing a strategy for an objective and mistaking an objective for a strategy. He would risk thus ending up with something quite different from the Japanese management or just its opposite.[18]

In a word, one should adopt or try to implement only Japanese management strategies whose goals are similar to his/her own. And if one wants to compete with the Japanese company, he/she should determine objectives that are either similar or superior to those of the Japanese company. If one's objectives and those of the Japanese company are similar, he/she may use Japanese management strategies to achieve them or different strategies should they be better than the Japanese ones. Superior objectives would for sure require superior strategies and perhaps the necessity of not using Japanese management strategies.

This approach wants to suggest one thing, that first before trying to apply Japanese management, one had better compare his/her management objectives and those of the Japanese company. And that does not compel him/her to use Japanese management strategies since different strategies may be aiming at the

---

18 I think that lifetime employment and lay-off should not be considered as principles and strategy respectively. They'd better be viewed as effects and priority interest principles, respectively

realization of the same goals. Japanese strategies should be used in the following case: 1) if they are superior in efficiency and effectiveness if one is aiming at those objectives for the first time.

Are Japanese management strategies more efficient or effective? If the answer is yes, the second step would be that of examining the principles governing your management environment, i.e., not only your company but also the shared management principle in your own industry and the surrounding industries. If they are similar to Japanese management, then you can be sure of succeeding in introducing Japanese management and later JIT system.

The importance of examining/knowing the Japanese management principle is double. First it prevents from trying to introduce a principle's effects. Under similar conditions, a principle would produce the same effects and it should be noted that effects occur necessarily when a principle is in action. Effects are consequence of applying a principle and effects are neither strategies nor objectives. Setting an effect as an objective would or might work once but that would not be permanent Second, the existence of principles governing the JCM system. It brings to the light the difficulty of introducing Japanese management in a different management context. Those principles are supra-company principles. Individual companies may endeavor to introduce them but if the general management environment shared by other companies does not supports them, they will not work.

Take for instance the commitment to one's employees and the training it involves. IBM is referred to as IBM university, i.e., it trains employees who would be hunted or poached by other companies. That is to say that the effort made by IBM to create a reliable work force and to develop an internal labor market force is being destroyed by the common management principle of relying on the external labor marketplace and market of skills, that is the non-commitment to one's employees.

And to understand the environment of Japanese management principles, it is necessary to go beyond the management framework and try to answer the question: Is Japanese management likely to work in a different culture? And that question is beyond the scope of that study. In fact, that question can not be answered in general tends by yes or no. Doing so would presuppose that world cultures are similar but they all differ from the Japanese culture only. The answer to that question needs a case-by-case study of different cultures of different countries in order to see the similarities and differences. And among the latter, one can distinguish differences that can be overcome from those that cannot. After only such a careful study case, one can then assess the probability of having JCM work in the considered foreign environment.

Summary, conclusion and perspectives

A. Japanese JIT and Japanese management

JIT was born and has developed inside the framework of the Japanese company management (JCM) with which it has developed strong and close ties. As I stated in a paper,

JIT is a sub-system of the Japanese management which is itself a sub-system of a more integrating system, i.e., the Japanese society.[1]

Figure A–1 Representation of JIT within JCM

Source: Adapted from Kupanhy, 1991

In order to reflect that inclusion of JIT within JCM (Figure A-1), the present thesis has adopted the following formal structure: JCM (chapters 1, 2) ==> JIT (chapters 3, 4, 5) ==> JCM (chapter 5, 6) The chapter five which, from the thesis' organization point of view serves as the bridge between JIT and management strategies is the one that deals in fact with the relationship between those two important spheres of the Japanese manufacturing company.

JIT and JCM have been approached as two consistent systems. Though JCM can be studied without any reference to JIT, the full understanding of the latter would however require also some knowledge of the former. The practical implications are:

---

1   L. Kupanhy, "Understanding JIT within the framework of the Japanese management system", 1991, p. 88

187

1. A complete implementation of JIT would presuppose a JCM-like environment

2. A partial implementation of JIT, i.e., JIT featuring only industrial engineering techniques, would work anywhere but it would yield only limited or partial results.

3. Due to the logical connection of the JCM features, it does not seem certain that an isolated management characteristic may work efficiently in a different milieu. Importing an element of JCM may imply the transfer of JCM itself or at least one of the management sub-systems the characteristic belongs to.

4. From a managerial point of view the Japanese management itself, in order to work in a different context, its underlying principles should not be contradicted by generally accepted management principles of the new environment.

The analysis of the survey results about the JCM has confirmed some doubt I had regarding the concept of the Japanese management in the current acceptation of the term. A tentative effort of clarification consisting in its re-formulation has been made so that its extension can cover the whole population of the Japanese industry.

That effort has led to the identification of features specific (1) to big corporations, (2) to small/medium enterprises and (3) to both large, small/mid-size corporations. Japanese big corporation management JBCM is thus made up of features (1) and (3) whereas features (2) and (3) are elements of Japanese small/mid-size company (JSCM). All those features, i.e., (1), (2) and (3), together make what l refer to as JCM

Thanks to the proposed prefiguration concept the dynamic aspect of forces within JCM as actualized through the relationship of its two sub-groups. I.e. JBCM and JSCM, has been brought to the light. JCM has been defined as made up of JSCM and JBCM while at the same time JSCM has been identified as featuring a structure that looks like a prefiguration of JBCM.

The same relationship of prefiguration has also been confirmed regarding the production methods. JIT has been seen as in its expansion among the small/mid-size companies and thus concerning the production methods of the Japanese company, the prefiguration of JIT has been qualified rather as emerging or taking form.

I have endeavored to provide an explanation of mine concerning the Japanese management and production methods because present theories left many questions answered. In fact, I do not feel quite satisfied with the contents of current concepts of the Japanese management/JIT since they erect rigid borders between management/production strategies of big corporations and those of small/

medium-size companies leading thus to problems that make it difficult to concili-
ate the theory with the reality.

On the other hand, Japanese management features are usually explained in the
framework of the Japanese culture and society, i.e., the latter is presented as the
foundation of the Japanese management system. At the same time, the same theories exclude the majority not only of the companies but also of the work force from
that group. I wondered if the overwhelming number of small and mid-size manufacturing companies and their work force that make up the majority of Japanese
workers are to be situated outside the Japanese society/culture so that their management looks neither influenced by nor founded in that society/culture.

Let me specifically point, among many others, to the following problems. It is
said that Japanese companies do not lay off employees because of the Japanese culture and tradition. Why do Japanese small corporations do so? Another dogma of
current theories about the Japanese management is that, because of their culture
and traditions, Japanese people identify themselves as members of a group and
when they join a company they will not leave it for another as do Westerners who
behave like mercenaries selling their knowledge to the best offer. Why do workers
of the small and mid-size company change companies? The cultural approach to
the Japanese management in spite of its enormous contribution to the understanding of the Japanese management is unable to provide a satisfactory answer to
those interrogations.

One has to remember that a theory is always a theory about something; it
refers to some reality it tries to reflect as accurately as possible. It is always an
attempt of a coherent explanation of some reality. Most theories on the Japanese
management seem to take a partial view of the reality of the Japanese corporations explaining its entrenchment in the Japanese society/culture. And the partial
view is presented as a global one. A have tried to fill the gap completing that
approach by the management-based approach. I thought explanations based on
the Japanese culture/society are correct but partial. They are not to be rejected but
should be completed by explanations that grasp the Japanese management from
within. Of course, this management-based explanation should also be considered
also as partial and complementary to other approaches to the Japanese company.

The interrogations above led to the hypothesis that there must be features of
the Japanese management that must apply to all Japanese companies regardless of
the company size if it is true that the social and cultural environments have some
decisive influences on the management system. Second, due to management constraints relating to the company size in terms of the work force, capital, share of
the market, etc., there must be features that are specific to big corporations only
and those which can apply only to small and mid-size manufacturing. The survey
data seem to have supported that double hypothesis.

B. Drawbacks of JIT and JCM

B.1. Inside the company

Though JIT is a marvelous production system that helps for sure improve the productivity, the quality of the product and that of the work[2] in a company that switches to it, the system is said to have given rise to some problems that can not be ignored.

At the company level, the waste of worker's waiting time (or time on hand) that JIT tries to reduce to zero, the multi-machine manning that requires a worker to supervise as many machine/processes as possible, the improvement regarding the reduction to the strict necessary minimum of the number of workers (hito-berashi) on the production line, the strictness of the tact time within which a worker should perform his different operations and the automatic stopping devices that can shut down a processing line in case a worker can not achieve his task within the prescribed cycle time make the life in some JIT companies hardly bearable.[3]

It is very unfortunate that though I have spent so many years studying JIT in its birthplace and visiting manufacturing companies, I have not, however, been able to verify those "accusations" against the JIT system. How could I since factory visits are always organized by managers whose main concern is to offer the public a nice image of their company? I think that only in an informal talk with line workers could one get an accurate idea about the real conditions prevailing on the shop floor.[4]

The efficiency of the JIT environment, i.e., Japanese management is out of question as one can realize it by a mere glance at the performance of the Japanese manufacturing company in general. That does not mean, however, that the system is 100% perfect.

---

2   See Detouzos, M. L., R. k. Lester and R. M. Solow, *Made in America. Regaining the competitive edge,* N. Y.: HerperPerennial, 1990; Schonberger, R. J., *Japanese manufacturing techniques. Nine Hidden lessons in simplicity,* N. Y.: The Free Press, 1982; Hall, R.W., *Zero inventories,* Homewood, Ill.: Dow Jones-Irwin, 1983; Shinohara, I., *NPS (New Production System): JIT crossing industry boundaries,* Cambridge, Mass., Productivity Press, 1988.

3   See Klein, J. A., "The human costs of manufacturing reform", *Harvard Business Review*, March April 1989, pp. 60–66. Fucini, J. J., and S. Fucini, *Working for the Japanese. Inside Mazda's American auto plant,* N.Y.: The Free Press, 1990

4   I met a friend's friend who works in a small rubber company that confirms to me JIT is really a hell for them and it is more hell for their subcontractors.

One of the elements that make the foreigner raise his eyebrows is the importance of the age which affects the salary on one hand and on the other hand its order of precedence over competence and its weight as regards the promotion process.

The Japanese explanation is that as one grows older, he gets more social responsibilities. Therefore, it is logical that his pay should be raised according to his age. But why should an un-married person get the same age-related raise of pay as his colleague of the same age who is married? Is there any relation between such a person's age and his social responsibilities? I think that does make much sense[5] since the social responsibilities of the employees are rewarded through different kinds of social allowances, such as family or housing allowances. Some scholars have pointed to the fact that the Japanese worker of the big corporation suffers from the lack of movement freedom due to the system of lifetime employment which runs counter the principle of free labor market. Odaka goes so far that for him lifetime employment for many employees means lifetime imprisonment:

The system of lifelong employment] appeals, however, only if the employee is able to find work that suits him within the corporation and the corporation is willing to let him pursue it (and very few employees are so lucky), or if the employee is content to be an ordinary salaried worker satisfied with whatever work, rank, and treatment the corporation assigns him. For such employees, lifelong employment was and is extremely beneficial, and no doubt they would agree heartily with the saying that "a big tree gives the best shelter". But to a person who cannot satisfy his personal needs in the company, who suffers from the feeling that he has gotten into the wrong place by mistake, and who would like to find a company that offers more congenial work and change careers before it is too late, lifelong employment can seem more like lifelong imprisonment—a cruel system whose motto is "Abandon all hope ye who enter here's. While lifelong employment contributes to employment stability, its is a source of considerable unhappiness for some people.[6]

Odaka is echoed by Sethi Namiki and Swanson who evaluate the human costs of the Javanese management:

And these costs may be substantial in terms of loss of individual freedom that may border on involuntary servitude, a rigid social structure, and sacrifice of other values individuals and groups may cherish but may be unable to exercise because of the intol-

---

5   I was regarded like some who has lost his sense when I tried, during a management workshop attended by mostly company managers (most of which are over forty), to express the same opinion!

6   K.Odaka, *Japanese management: a forward-looking analysis,* Tokyo: Asian Productivity Organization, 1986. p. 63–64

erance of a system that is structurally imposed and from which escape may be all but impossible.[7]

Another disadvantage of the Japanese management is identified as relating to the seniority-based system of rewards which, according to Odaka is advantageous for mediocre people.

The seniority-based hierarchy offers very real benefits to the employees of a large corporation. Even mediocre employees, as long as they obey orders and are patient, can expect to be promoted and accorded better treatment almost automatically as they pile up more years of service with the company ... The result is that the system creates a large number of outwardly efficient but inwardly mediocre and lazy company men who are "never idle, never late, but never actually working". This is detrimental not only to the employee with outstanding abilities but also to the company as a whole.[8]

At last, let me point to the following contradiction. The commitment of the Japanese company to its full-time employees means in fact non-commitment to temporary employees, female employees, part-time employees, seasonal or daily laborers who are regarded as the cushion work force, i.e., those to be laid off any the company goes through some hardships.

The cushion work force is not covered by the principle of the fair distributions of the company profits since they get less pay, less or no bonus at all even though their work load is the same as that of the privileged full-time regular workers.

B.2. Within the manufacturing sector

Within the industry, lifetime commitment to big corporation's employees implies non-lifetime in the small manufacturing firms that are sub-contractors. The promotion from within that characterizes big corporations sometimes means non-promotion from within in the subsidiaries and supplier companies where important retiree from big corporations are parachuted as senior managers and/or members of the board of the subcontractors.[9]

Concerning the manufacturing methods, it is worth noting that the non stock production for example practiced by big JIT companies and the just-in-time deliveries of purchased products, sometimes if not very often, forces in fact the small subcontracting companies to use the stock-based production in order to deliver parts at the earliest within hours from the moment they get an order from the

---

7   Sethi, S. P., N. Namiki and C. L. Swanson, *The false promise of the Japanese miracle. Illusions and realities of the Japanese management system,* Boston: Pitman, 1984, p.48

8   Odaka, K., *Japanese management,* 1986, p. 65–66

9   See Clark, op. cit.

parent JIT company. That has led some supplier companies to run stock houses situated close to the parent company site.[10]

C. Perspectives

Those few negative points do not throw any shadow on JIT and its immediate management environment i.e., JCM nor do they shake their overall economic efficiency. They should be considered as seeds for improvement that can help better the two systems.[11]

As far as the Japanese company continues to sustain an economy that maintains Japan at the rank of an economic super-power, JIT and JCM will not be abandoned overnight in Japan. Slight changes consisting in the systems improvement and/or efforts of their adaptation to new situations remain however possible. And foreign scholars and businessmen will continue looking to the Japanese company as a model and a source of inspiration.

In order to emulate and/or compete with the Japanese company, the foreign company should understand the tools it uses to produce and how it uses them. To understand the outstanding performance of the Japanese company, a deep look into its management system is necessary. After mastering those management and production techniques one can examine the possibility of their application elsewhere. The present research has tried to shed some light on those aspects of the Japanese manufacturing company.

---

10 See Nikkei Komyunikeishon (ed.), *Muzaiko keiei. "Joteku" ga shijo WO seisu,* Tokyo Nihon Keizai Shinhun-sha, 1987; Sato, Y., *Toyota gurupu no senryaku to jissho bunseki,* Tokyo: Hakutoshobo, 1988

11 At JR, only competence prevails. See Nihon Keizai Shimbu-sha (ed.), *Terase de yomu nihon no keiei.* Tokyo: Nihon Keizai Shimbun-sha, 1989

References

Abegglen, J. C. *The Japanese factory: Aspects of its social organization*. N.Y.: Arno Press, 1979. (Reprint, originally published by Glencoe, Ill: The Free Press, 1958)

Abegglen. J. C. & G. Stalk, Jr. *Kaisha, the Japanese corporation*. Tokyo: Ch. E. Tuttle Co., 1987 Abegglen, J.C. Nihon no kigyo shakai. Kyoto: Koyoshobo. 1989

Abernathy, W. J. and K. B. Clark and A. M. Kantrow. "The new Industrial competitor," *Harvard Business Review*, Sept/Oct. 1981, pp. 68–81

Ahsanduddin, A., "Survey identifies critical factors in successful implementation of Just-in-time purchasing techniques," *Industrial Engineering*, October 1986, pp. 44–50

Aichi Rodo Mondai Kenkyu-sho, ed. *Toyota gurupu no shin-senryaku*. Tokyo: Shin Nihon Shuppan-sha, 1 990

Anderson, Ch. A. "Corporate directors in Japan," *Harvard Business Review*, May/June 1984, pp.30–38

Arnesen, P. J., ed. *The Japanese : Phase 2* (Michigan Papers in Japanese studies, no 15), Ann Arbor, Mich.: Center for Japanese Studies, The University of Michigan, 1987

Benedict, R. *The Chrysanthemum and the sword: Patterns of Japanese culture*. Tokyo: Ch. E. Tuttle, 1 954 (47th printing, 1992)

Best, M. H. *The new competition: Institutions of industrial restructuring*. Cambridge: Polity Press, 1990

Buffa, E. S.and R. K. Sarin. *Modern production/operation management*. N.Y.: John Wiley & Sons, 1987

Ch. E. Tuttle, 1979 Vogel, E. F. *Comeback: Case by case building the resurgence of American business*. Tokyo: Ch. E. Tuttle, 1985 Vogel, E. F. Japan as no 1: Lessons for America. Tokyo Ch. E. Tuttle, 1980

Christopher, R. C. *Second to none: American companies in Japan*. Tokyo: Ch. E. Tuttle, 1986

Chusho Kigyo Cho, ed. *Zu de miru chusho kigyo hakusho*. Tokyo. Okura-Sho, 1992

Clark, R. *The Japanese company*. Tokyo: Ch. E. Tuttle Co., 1987

Clegg, W. H. "Operator/machine studies Technique reduces set-up time, implements JIT", *Industrial Engineering*, October 1986, pp. 54–53

Cole, R. E., ed. *The American automobile industry: Rebirth or requiem?* (Michigan Papers in Japanese studies, no. 15). Ann Arbor, Mich: Center for Japanese Studies, The University of Michigan, 1984

Das, P. K. "Simplifying material handling system cuts losses due to damages 70%", *Industrial Engineering*, July 1988, pp. 47–51

Detouzos, M. L., R. K. Lester, and R. M. Solow. *Made in America: Regaining the competitive edge*. N. Y.: HarperPerennial, 1990

Discasali, R. L. "Job shops can use repetitive manufacturing method to facilitate just-in-time production", *Industrial Engineering*, June 1986, pp. 48–52

Dore, R. P. and M. Sako. *How the Japanese learn to work*. London: Routledge, 1989

Drucker, P. F. "Behind Japan's success", *Harvard Business Review*, January/February 1981, pp. 83–90

Edosomwan, P. *The effective executive*. N. Y.: Harper & Row, 1985

Engwall, R. L. "The expanding role of IEs in manufacturing", *Industrial Engineering*, June 1989, pp. 52–53

Feather, J. J. and K. F. Cross. "Workflow analysis: Just-in-time techniques simplify, administrative process in paper work operation", *Industrial Engineering*, January 1988, pp. 32–40

Ford, H. *Today and tomorrow*. Cambridge, Mass.: Productivity Press, 1988 (Reprint. Originally published by Garden City, NY: Doubleday, Page & Co., 1926)

Friedman, D. *The misunderstood miracle: industrial development and political change in Japan*. Ithaca, N.Y.: Cornell University Press, 1988

Fucini, J. J. and S. Fucini. *Working for the Japanese: Inside Mazda's American auto plant*. N.Y.: The Free Press, 1990

Garvin, D. A. "Quality on the line", *Harvard Business Review*, September/October, 1983, pp.65–75

Gitlow, H. S. and P. T. Hertz. "Product defects and productivity". *Harvard Business Review*, September/October, 1983, pp. 131–141

Grandori, A. *Perspective on organization theory*. Cambridge, MA: Ballinger Publishing Co., 1987

Grieco, Jr., P. L., W. G. Michael and J. W. Claunch. *Just-in-time purchasing: In pursuit of excellence*. Plantsville, Ct: PT Publication, 1988

Gross J. M. and K; R. Mcinnis. *Kanban made simple: demystifying and applying Toyota's legendary manufacturing process*. NY: AMACOM, 2003

Haas, E. A. "Break-through manufacturing", *Harvard Business Review*, March/April 1987, pp. 75–81

Hall R. W. "Synchro MRP: combining kanban and MRP, the Yamaha PYMAC system". Y. Monden, ed. *Applying just in time: the American/Japanese experience*, Norcross, Ga: Institute of Industrial Engineers, 1986, pp. 18–31

Hall R. W. *Zero inventories*. Homewood, Ill.: Dow Jones-Irwin, 1983

Hall, E. T. and M. R. Hall, *Hidden differences: Doing business with the Japanese*. N.Y.: Anchor Books, 1987

Harada, K. *Chiisana kaisha no jozuna hito no torikata*. Tokyo: Oesu, 1990

Hayes, R. H. "Why Japanese factories work". *Harvard Business Review*, July/August, 1981, pp. 57–66

Hazama, H *Nihonteki keiei: Shudan shugi no kozai*. Tokyo: Keizai Shimbun-sha, 1971

Hazama, H. *Nihonteki keiei no keifu*. Tokyo: Nihon Noritsu Kyokai, 1963

Hill, I. D. "Modern manufacturing techniques require flexible approqch to facility plqnning", *Industrial Engineering*, May 1984, pp. 86–93

Hilton, Ch. B. "Japanese management: Clockwork or chrysanthemum: an American perspective", *Osaka City University Business Review*, No. 3, pp. 1991–1992, pp. 67–73

Hirai, T. *Toyota cho-koshueki keiei no hasso.* Tokyo: Paru Shuppan, 1990

Hirano, Hirayuki. *Kanban to me de miru kanri.* Tokyo: Nikkan Kogyo, 2001

Howard, R. "Can small business help countries compete?", *Harvard Business Review*, November/December 1990, pp.88–103

Huddleston, Jr., J. N. *Gaijin kaisha: Running a foreign business in Japan.* Tokyo: Ch. E. Tuttle, 1990

Ibrahim, N. and A. D. Nicoll. "Project planning network is integrated plan for implementing just-in-time", *Industrial Engineering*, October 1987, pp. 50–55

Iizuka, Y. and Y. Monden. "Mechanism of supplier's response to the kanban system". Y. Monden, ed. *Applying just in time: the American/Japanese experience*, Norcross, Ga: Industrial Engineering and Management Press, 1986, pp. 69–78

Imai, M. *Kaizen: The key to Japan's competitive success.* Singapore: McGraw-Hill, 1991

Itami, H., T. Kagono, T. Kobayashi, K. Sakakibara and M. Ito. *Kyoso to kakushin: jidosha sangyo nokgyo seicho.* Tokyo: Toyo Keizai Shinpo-sha, 1988

Johansen, P. and K. J. McGuire. "A lesson in SMED with Shigeo Shingo", *Industrial Engineering*, October, 1986, pp. 26–33

Jonstone, J. and J. Koenig. "Flow reviews systematically improve manufacturing process logistics", *Industrial Engineering*, 1987, pp. 40–44

Jung, H. F. how to do business with the Japanese. Tokyo: The Japan Times, 1986

Kachur, R. G. "Electronics firm combines plan move with switch to JIT manufacturing", *Harvard Business Review*, March 1989, pp. 44–48

Kagono, T., and Kansai Productivity Center/Kansai seisansei honbu, eds. *How Japanese companies work/Midoru ga kaita nihonteki keiei.* Tokyo: Nihon Keizai Shimbun-sha, 1984

Kagono, T., I. Nonaka, K. Sakakibara and A. Okumura. *Strategic vs. evolutionary management.* A U.S.- Japan comparison of strategy and organization. N. Y.: Elsevier Science Publishers B. V. 1985

Karatsu, H. *Nihonteki keiei ha shinazu.* Tokyo: PHP, 1993

Kimura, O. and H. Terada. "Design and analysis of Pull-System, a method of multi-stage production control". Y. Monden. *Toyota Production system: A practical approach to production management*, Atlanta/Norcross, GA: *Industrial Engineering* and Management Press, 1983, pp. 219–232

Klein, J. A. "The human costs of manufacturing reform", *Harvard Business Review*, March/April 1989, pp. 60–66

Kobayashi, N. *Toyta no dai-jikken: Sore ha jinjikakumei kara hajimatta*. Tokyo: Shoden-sha, 1990

Konz, S. "Quality circles: Japanese success story", *Industrial Engineering*, October 1979, pp. 24–27

Kosei Torihiki Iinkai Jimu Kyoku, ed. *Nihon no 6-dai kigyo shudan: Sono soshiki to kodo*. Tokyo: Toyo Keizai Shinpo-sha, 1992

Kupanhy, L. "Does the Japanese small and mid-size manufacturing use JIT? A survey-based study", *Osaka City University Business Review*. No. 4, 1993- 1994

Kupanhy, L. "Japanese management in the small and mid-size manufacturing: a survey", *Keiei Kenkyu*, Vol. 43, No. 3, 1992, pp. 47–61

Kupanhy, L. "Understanding JIT within the framework of the Japanese management system", *Keiei Kenkyu*, Vol. 42, No. 2, 1991, pp. 87–112

Kupanhy, L. *Just-in-time manufacturing management: the case of Toyota Production System*. (Thesis for the Master's Degree in Economics, Wakayama University, Japan, 1988, unpublished)

Kupanhy. L. "Inefficiencies of the Zairian manufacturing", *Osaka Shidai Ronshu*, No. 69, December 1992, pp. 29–43

Lawler III, E. E. and S. Mohrman. "Quality circles after the fad", *Harvard Business Review*, January/February 1985, pp. 65–71

Lemon, E. J. *Beginning logic*. Ontario: T. Nelson and Sons, Ltd., 1965

Lesnet, D. E. "Facility found that means of implementing quick die changes were already at hand", *Industrial Engineering*, November 1983, pp. 50–59

Liker, J. K. *The Toyota way: 14 management principles*. NY: McGraw-Hill, 2004

Low, A. *Zen and creative management*. Tokyo: Ch. E. Tuttle, 1992

Lu, D. J. *Inside corporate Japan: The art of fumble-free management*, Tokyo: Ch. E. Tuttle, 1989

March, R. M., *Working for a Japanese company: Insights into the multicultural workplace*. Tokyo: Kodansha International, 1992

Matsushita, K. *Quest for prosperity: the life of a Japanese industrialist*, Kyoto: PHP, 1988

McMillan. Ch. J.. *The Japanese industrial system*, Berlin: Walter de Gruyter, 1985 (second revised edition)

Monden, Y. "What makes the Toyota Production System really tick", *Industrial Engineering*, January 1981, pp. 36–46

Monden, Y. *Applying just in time: the American/Japanese experience*. Norcross, GA: *Industrial Engineering* and Management Press, 1986

Monden, Y. Toyota *Management System: Linking the seven key functional areas*, Portland, Oregon: Productivity Press, 1993

Monden, Y. *Toyota Production System: An integrated approach to Just-in-time*, 3rd ed. Norcross, GA: Engineering & Management Press, 1998

Monden, Y., R. Shibakawa, S. Takayanagi and T. Nagao, eds. *Innovation in management: The Japanese corporation*. Norcross, Georgia: Industrial Engineering and Management Press, 1985

Monden, Y., *Toyota Production System: practical approach to production management*. Atlanta/Norcross, Georgia: Industrial Engineering and Management Press, 1983

Monden. Y. "How Toyota shortened supply lot production time, waiting time and conveyance time". *Industrial Engineering*, September 1981, pp. 22–30

Monden. Y. "Smoothed production lets Toyota adapt to demand changes and reduce inventory", *Industrial Engineering*, August 1981, pp. 42–51

Monden. Y., "Adaptable kanban system helps Toyota maintain just-in-time production", *Industrial Engineering*, May 1981. pp. 29–46

Mori, H. *TQC no chishiki*. Nihon Keizai Shimbun-sha, 1983

Morita, A., (with E. M. Reingold and M. Shimomura). *Made in Japan: Akio Morita and Sony*. Glasglow: Fontana?Collins, 1987

Muller D. M. "Profitability = productivity + price recovery", *Harvard Business Review*, May/June 1984, pp. 145–153

Nakagawa, K., ed. *Kigyo keiei no rekishiteki kennkyu*. Tokyo: Iwanami Shoten, 1990

Nakane, J. and R. W. Hall, "Management specs for stockless production", *Harvard Business Review*, May/June 1983, pp. 84–91

Namiki, t. *Kojo kanri no chishiki*. Tokyo: Nihon Keizai Shimbun-sha, 1972

Nemoto, M. *Total quality control for management: strategies and techniques from Toyota and Toyoda Gosei*. Englewood Cliff, NJ: PrenticeHall, 1987

Nevins, T.J. *Labor pains and the gaijin boss: Hiring, management and firing the Japanese*. Tokyo: The Japan Times, 1985

Nihon Keizai Shimbun-sha, ed. *Terase de yomu nihon no keiei*. Tokyo: Nihon Keizai Shimbun-sha, 1989

Nihon Noritsu Kyokai, ed. (shinpan zoho: Monden Y.). *Toyota no genba kanri: Kanban hoshiki no tadashii susumekata,*. Tokyo: Nihon Noritsu Kyokai Manejimento Senta, 1986

Nikkei Komyunikeishon, ed. *Muzaiko keiei: 'Jo teku' ga shijo wo seisu* (制す). Tokyo: Nihon Keizai Shimbun-sha, 1987

Odaka, K. *Japanese management: a forward-looking analysis*. Tokyo: Asian Productivity Organization, 1986

Ogawa, E. *Gendai no seisan kanri*. Tokyo: Nihon Keizai Shimbun-sha, 1982

Ohmae, K. *The mind of the strategist*. N.Y.: Penguin Books, 1983

Ohno T., and Y. Monden, eds. *Toyota seisan hoshiki no shin-tenkai*. Tokyo; Nihon Noritsu Kyokai, 1983

Ohno, T. Ohno *Taiichi no genba keiei*. Tokyo: Nihon Noritsu Kyokai, 1982

Ohno, T. *Toyota Production System: Beyond large-scale production*. Cambridge, MA: Productivity Press, 1988

Ohno, T. *Toyota seisan hoshiki: Datsu-kibo no keiei wo mezashite*. Tokyo: Daiyamondo-sha, 1978

Ohno, T. *Workplace management* (translation by A. P. Dillon), Cambridge, MA: Productivity Press, 1988

Ohsono, T. *Hito me de wakaru kigyo keiretsu to sakai chizu: Nihon no sangyokai ni harereta tate-ito to yoko-ito wo yomu.* Tokyo: Nihon Jitsugyo Shuppan, 1991

Ohtsuki, N. *Toyota no shin kanban hoshiki.* Tokyo: Chkei Shuppan, 1985

Oliver N., and B. Wilkingson. *The Japanization of British industry,* N.Y.: Basil Blackwell, 1988

Ouchi, W. G. *Theory Z: How American business can meet the Japanese challenge.* N.Y.: Avon books, 1982

Pascale, R. T. & A. G. Athos. *The art of -the Japanese management: application for American executives.* N.Y.: Warner Books, inc, 1982

Prestowitz. Jr.,. C. V. *Trading places: How America allowed Japan to take the lead.* Tokyo: Ch. E. Tuttle, 1988

Production at Toyota: Our basic philosophy (The 5th in a series published by by Toyota Motor corporation), Toyota Motor Corporation, (without date)

Reddy, J.. and A. Berger. "Three essentials of product quality", *Harvard Business Review* July/August 1983, pp. 153–159

Robinson, R. G., and D. M. Schroeder, "Training, continuous improvement and human relations: the U.S. TWI Programs and the Japanese management style", *California Management Review*, Vol. 35, No. 2, winter 1993, pp.35–57

Sakai. K. "The feudal world of Japanese manufacturing, *Harvard Business Review*, November/December, 1990, pp. 38–51

Sasaki, O. *Hinsitsu kanri no chishiki.* Tokyo: Nihon Keizai Shimbun-sha, 1972

Sato, Y. *Toyota gurupu no senryaku to jissho benseki.* Tokyo: Hakutoshobo, 1988

Schonberger, R. J. "Just-in-time production system: replacing complexity with simplicity in manufacturing management", *Industrial Engineering*, October, 1984, pp. 52–63

Schonberger, R. J. *Japanese manufacturing techniques: Nine hidden lessons in simplicity.* N. Y.: The Free Press, 1982

Schonberger, R. J. "Plant layout becomes product-oriented with cellular, just-in-time production concepts", *Industrial Engineering.* November, 1983, PP. 66–71

Schonberger, R. J. "Production workers bear mayor quality responsibility in Japanese industry", *Industrial Engineering*, December 1982, pp. 34–40

Schonberger, R. J. *World class manufacturing: The lessons of simplicity applied.* N. Y.: The Free Press, 1986

Schonberger, R. J., "Frugal manufacturing", *Harvard Business Review*, September/ October 1987, pp. 95–100

Schonberger, R. J., "The human side of kanban", *Industrial Engineering*, August 1993, pp. 34–36

Schroer, B. J., J. T. Black and Shou Xiang Zhang, "Microcomputer analyzes 2- card kanban system for 'Just-in-time' small batch production" *Industrial Engineering* June 1984, pp. 54–64

Sepehri, M. "How kanban system is used in American Toyota Motor facility", *Industrial Engineering*, February 1985, pp. 50–56

Sepehri, M. "Quality and inventory control go hand i hand at Hewlett- Packard's computer system division", *Industrial Engineering*, February 1986, pp. 54–61 (?)

Sepehri, M. "Competition requires management to focus attention on manufacturing", *Industrial Engineering*, May 1987, pp. 6–8

Sepehri, M. "Manufacturing revitalization at Harley-Davidson Motor Co.", *.Industrial Engineering*, August 1987, pp. 87–93

Sepehri, M., "Car manufacturing joint venture tests feasibility of Toyota method in U.S.", *Industrial Engineering*, March 1986, pp. 34–41

Sethi, S. P., N. Namiki, and C. L. Swanson. *The false promise of the Japanese miracle: Illusions and realities of the Japanese management system.* Boston: Pitman, 1984

Shapiro. H. J. and T. Cosenza, *Reviving Industry in America: Japanese influences on manufacturing and the service sector.* Cambridge, MA: Ballinger, 1987

Shingo, Sh. *Non-stock Production: The Shingo system for continuous improvement.* Cambridge, MA: Productivity Press,1988

Shingo, Sh. *Study of Toyota Production System from Industrial Engineering point of view.* Tokyo: Japan Management Association, 1981

Shinohara, I. *NPS (New Production System): JIT crossing industry boundaries.* Cambridge, MA: Productivity Press, 1988

Shinohara, I. *NPS no kiseki: be-ru wo nuida seisan hoshiki.* Tokyo: Toyo Keizai Shimbun-sha, 1985

Shinohara, I. *NPS seisan hoshiki: fumetsu no keiei.* Tokyo: Toyo Keizai Shimbun-sha, 1989

Siebeneicher, Robert 2nd, '"Education, trained employees are key to success for today's business", *Industrial Engineering*, March 1987, pp. 44–55

Simers, D., John Priest and Jack Gary, "Just-in-time techniques in process manufacturing reduced lead time, cost; raise productivity, quality", *Industrial Engineering*, January 1989, pp. 19–23

Skinner, W. *Manufacturing: the formidable competitive weapon.* N Y.: John Wiley & Sons, 1985

Smltka, M. J. *Competitive ties: Subcontracting in the Japanese automotive industry.* N.Y.:Columbia University Press, 1991

Takagi, T. *Toyota kara kacho ga kieta: Mono, kane kara hito no katsu seika no jidai he.* Tokyo: Gomashobo, 1990

Tamagara, T. *Toyota hoshiki ni miru shisutemu sai kochiku: Kigyo kakushin wo ninau kanrisha no tame ni.* Tokyo: Paru Shuppan, 1988

Terasawa, H. *OJT no jissai.* Tokyo: Nihon Keizai Shimbun-sha, 1989

Thian, H, *Setting up and operating a business in Japan.* Tokyo: Ch E Tuttle, 1988

Uchino, T. and J. C. Abegglen, eds. *Tenki ni tatsu nihongata kigyo keiei.* Tokyo: Chuo Keizai Sha, 1988

Ueda, T. *Sho-shudan katsudo no tebiki.* Tokyo: Nihon Keizai Shimbun-sha,1980

Uemura, Sh. "Keiei senryaku/keiei soshiki no tenkai no nihonteki keiei", *Minshuka/gorika kara johoka/kokusaika he.* Tokyo: Bunshindo, 1991, pp. 120–153

Uemura, Sh. *Nihonteki keiei soshiki.* Tokyo: Bunshindo, 1993

Uemura, Sh. *Soshiki no riron to nihonteki keiei.* Tokyo: Bunshindo, 1982

Uemura, Sh. "The Japanese way of management: its characteristics, current practices, and future perspectives, *Osaka Citv University Business Review*, No 2, 1989, pp. 15–26

Urabe, K "Innovation and the Japanese management system", K. Urabe, J. Child, and T Kagono, eds. *Innovation and management: International comparison.* Berlin: Walter de Gruyter, 1988, pp. 3–25 Vogel, E. F. ed. Modern Japanese organization and decision-making. Tokyo.

Wakamatsu, Yoshihito. *Toyota seisan ryoku: monozukuri kyukyoku no chie.* Tokyo: Dayamondo, 2001

Wakamatsu, Yoshihito. *Toyota shiki kyukyoku no jissen.* Tokyo: Dayamondo 2002

Walleigh, R. and M. Sepehri, "H-P Division programs reduce cycle times, set stage for ongoing process improvements", *Industrial Engineering,* March 1986, pp. 74–81

Walleigh, R. C., "Getting things done: what's your excuse for not using JIT?" *Harvard Business Review*, March/april, 1986, pp. 38–54

Waters, D., *21st century management: keeping ahead of the Japanese and Chinese.* N.Y.: 1991 Weis, A., "Simple truths of Japanese manufacturing, *Harvard Business Review*, July/August, 1984, pp. 119–125

Wheelwright, S. C., "Japan- where operations really are strategic", *Harvard Business Review*, July/August 1981, pp. 67–74

Womack, J. P., D.T Jones and D. Roos. *The machine that changed the world. The story of lean production.* N.Y. Harper Perennial, 1991

Womack, J.P. and D. T; Jones Lean thinking: Banish waste and create wealth in your company. NY: Simon & Schuster, 1996

Yang, Ch. Y. "Demistifying Japanese management practices, *Harvard Business Review.* Nov./December, 1984, pp. 172–184

Yashiro, M., *Nihon no keiei/America no keiei.* Tokyo: Nihon Keizai Shimbun-sha 1992

Zimmerman, M., *How to do business with the Japanese.* Tokyo: Ch. E. Tuttle, 1987

Index

978-0-595-45438-9
0-595-45438-0